Crystal Research and Technology

2

Journal of Experimental and Industrial Crystallography

Zeitschrift für experimentelle und technische Kristallographie

Volume 23
1988
Number 2 February

THEORIA CUM PRAXI

Akademie-Verlag · Berlin

ISSN 0232-1300 Cryst. Res. Technol., Berlin 23 (1988) 2, 141–296, K29–K50

Directions to the author

In the journal "Crystal Research and Technology – Kristall und Technik" as a rule the following categories of papers will be printed: original papers, review articles, short notes, and book reviews, **preferentially in English.** Complete information necessary for type-setting and a "Short Title" should be added.

The manuscript should not exceed 10 pages and be type-written on one side only in the format A 4 (21 × 29,7 cm) on white ink-proof paper, with a two-lined distance and a left-hand free edge of 3 cm, and submitted in 2 copies. All sheets of the paper belonging to the contribution are continuously to be numbered.

The head of the manuscript should be arranged in this manner: initials and name of the author; institution, and title of the paper. Then two abstracts should follow, one in the original language and the second in German, English, French, or Russian, respectively.

Formulae, signs and symbols, letters etc. being wanting on the type-writer, should be distinctly written with penholder. O as zero is to be underlined, the capital O is not. Greek letters should be underlined with red ink.

Paragraphs should be marked by indenting of at least 3 touches.

Concerning the marking of the manuscript should be noted:
headings should be underlined, sub-headings not, and indented by 9 touches. The titles of tables are not to be underlined. Other markings in the text will be done by the editorial staff. They mean:

light-red 〰〰〰 = small capitals
blue 〰〰〰 = in italics

light-red | perpendicular straight red line on the left edge = small printing

Tables are to be added on separate sheets of paper at the end of the contribution.

Foot notes should be continuously written on a separate sheet of paper.

Figures (Fig.) – drawings and photographs – should be not more than 8 and submitted (2 copies) in a special envelope: photographs as white high-gloss prints (inscription on a transparent covering sheet) and drawings as distinct sketches. On the back of the submitted figures the reduction scale in tenth (9/10, 8/10, 5/10 e.g.), author's name, short title of the paper, and the number of the figures should be given.

Captions for the figures should be submitted on a separate sheet of paper.

References should be written without numbers in alphabetical order of the names at the end of the contribution. Cited names in the text stand for the references. The references should contain the following items, for example:

for books:	RAST, K.: Physikalisch-chemische Rechnungen, Berlin 1958
for journals:	KRAUSE, H. N., MÜLLER, K.: Cryst. Res. Technol. **16** (1981) 70

Add, please, the short title, running title, about 6 keywords, and the academic grades of the authors.
The authors will be sent two galley proofs of review articles and original papers. Proof reading of short notes and book reviews will be done by the editorial staff.

For each of these papers a total of 50 reprints is available free of charge. Requests for additional maximum 200 reprints (against payment) should be given together with the submitted manuscript.

Founded 1966 as
Kristall und Technik

Crystal Research and Technology 2

VOLUME 23 · 1988 · NUMBER 2

Journal of Experimental and Industrial Crystallography

THEORIA CUM PRAXI
Akademie-Verlag · Berlin

Papers should be directed to the editor's office
DDR-7030 Leipzig, Scharnhorststraße 20,
or to the Editor-in-Chief Prof. Dr. Hermann Neels,
DDR-7254 Machern, Zeititzer Weg 25,
or to one of the members of the Editorial Board or Advisory Board.

Orders can be sent

- in the GDR:
  to the Postzeitungsvertrieb or to the Akademie-Verlag Berlin, Leipziger Straße 3–4, DDR-1086 Berlin;
- in the other socialist countries: to a book-shop for foreign language literature or to the competent news-distributing agency;
- in the FRG and Berlin (West): to a book-shop or to the wholesale distributing agency Kunst und Wissen, Erich Bieber OHG, Wilhelmstr. 4–6, D-7000 Stuttgart 1;
- in the other Western European countries: to Kunst und Wissen, Erich Bieber GmbH, General Wille-Str. 4, CH-8002 Zürich;
- in USA and Canada: to the VCH Publishers, Inc., 303 NW 12th Avenue, Deerfield Beach, FL 33442-1788, USA;
- in other countries; to the international book- and journal-selling trade, to Buchexport, Volkseigener Außenhandelsbetrieb der DDR, Postfach 160, DDR-7010 Leipzig, or to the Akademie-Verlag Berlin, Leipziger Straße 3–4, DDR-1086 Berlin.

Journal „Crystal Research and Technology"

Herausgeber: Im Auftrage des Akademie-Verlages Berlin herausgegeben von Prof. Dr. Hermann Neels.

Verlag: Akademie-Verlag Berlin, Leipziger Str. 3–4, DDR-1086 Berlin; Fernruf: 2236221 oder 2236229; Telex-Nr.: 114420; Bank: Staatsbank der DDR, Berlin, Kto.-Nr. 6836-26-20712.

Chefredakteur: Prof. Dr. Hermann Neels, Zeititzer Weg 25, DDR-7254 Machern; Fernruf: Brandis 3442. Anschrift der Redaktion: Scharnhorststr. 20, DDR-7030 Leipzig; Telefon: 310502.

Veröffentlicht unter der Lizenznummer 1526 des Presseamtes beim Vorsitzenden des Ministerrates der Deutschen Demokratischen Republik.

Gesamtherstellung: VEB Druckerei „Thomas Müntzer", DDR-5820 Bad Langensalza.

Erscheinungsweise: Das Journal „Crystal Research and Technology" erscheint monatlich. Die 12 Hefte eines Jahrganges bilden einen Band. Bezugspreis je Band 480,– DM zuzüglich Versandspesen; Preis je Heft 40,– DM. Der gültige Jahresbezugspreis für die DDR ist der Postzeitungsliste zu entnehmen.

Bestellnummer dieses Heftes: 1082/23/2.

AN (EDV) 35830
02800

Cryst. Res. Technol.	23	1988	2	259–261

Original Papers

L. N. Aleksandrov, A. N. Kogan, R. V. Bochkova, N. P. Tikhonova

Institute of Semiconductor Physics, Siberian Branch of the USSR Academy of Sciences, Novosibirsk, and
Mordovian State University, Saransk

Study of the Chemical Reactions' Influence on Semiconductor Films Structure Formation by the Monte-Carlo Method

The investigation by the Monte-Carlo method of the growth of the silicon epitaxial film at a chloride CVD system has allowed to find out the composition of adsorption layer, the micromechanism of the reactions of Si atoms building-in into the growing crystalline layer and the growth conditions influence on the growth rate and film surface roughness. The change of adsorptive layer composition in the system $SiCl_4$–HCl–H_2 (fraction of ada-atoms, silicon atoms built-in a crystal and molecules $SiCl_2$) depending on temperature has been determined. The change of silicon film growth rate depending on temperature and concentration change of SiH_2Cl_2 has been established and the contribution of growth mechanism (with participation of adatom, silicon atoms and molecules $SiCl_2$) into the total rate of film growth has been shown.

Проведенное методом Монте-Карло изучение роста эпитаксиальных пленок кремния в хлоридной газотранспортной системе позволило выяснить состав адсорбционного слоя, микромеханизм реакций встраивания атомов кремния в растущий кристаллический слой и влияние условий роста на скорость роста и шероховатость поверхности пленок. Установлено изменение с температурой состава адсорбционного слоя в системе $SiCl_4$–HCl–H_2: доля адатомов и в троенных вскристалл атомов кремния и молекул $SiCl_2$. Установлено изменение скорости роста пленки кремния с температурой и изменением концентрации SiH_2Cl_2 и показан вклад в обшую скорость роста пленки механизмов роста с участием адатомов и атомов кремния, и молекул $SiCl_2$.

1. Introduction

The possibility of computer simulation of the semiconductor films growth processes has been shown in a number of papers (Aleksandrov et al. 1985 a, b, c). The film production from a molecular beam has been considered in these papers.

The goal of the present work is to study the growth peculiarities of the semiconductor film CVD method by the computer statistic simulation method. The crystallization by chemical reactions is, by Givargizov, a peculiar crystallization, a chemical one, which has the important advantages in the growth control in comparison with the other methods. The chemical reactions not only supply the growth surfaces with a material, but their interestitial products can directly be concerned in the crystallization process. Owing to a chemical reaction the crystal growth assumes the features of a self-regulating process: by its parameter change, the conditions close to equilibrium maintain automatically and the growth error correction occurs.

However, the mechanism of growth process from the vapour phase with a chemical reaction is sufficiently complicated than the crystal growth obtaining by a condensation from vapour phase. The film growing by the CVD method is connected with many processes: the diffusion in a gas phase and on a surface, a chemical reaction and crystallization. The problem of transport kinetics investigation consists in these processes division, in the determination of their contributions into the total growth rate.

The crystal growth investigation in gas-transport systems must solve three problems:

1. The determination of a gas phase composition of the growing crystal.
2. The study of crystal growth mechanism and kinetics from the gas phase of a given composition.
3. The study of an adsorption layer composition and structure and the study of a crystallization surface.

1. The first problem is a gas-dynamic and thermodynamic one. It is necessary for the obtaining of any reaction equation to identify all chemical compounds and simple materials, which are concerned in the reaction. The results of thermodynamic calculations are used for promary determination of a gas-phase composition. The lack of information is caused by the deficiency in the initial thermodynamic characteristics.

2. The transport mechanisms of the materials should be considered for the determination of optimal conditions and increase of growth process efficiency. In this case the main problem is the determination of a stage which limits the process rate (Mass-transport or heterogeneous reaction kinetics). This problem can be solved by a phenomenological consideration or by a microscopic one.

The phenomenological approach allows to obtain the data of total growth rate and limiting stages of the process macromechanism. The process rate is determined by the rate of its slowest stage. It is necessary to determine the dependences of different processes on the supersaturation, temperature or flow rate in order to find out what stage determines a process rate. The data concerning process which determines the film growth rate at chemical transport reactions are discrepant (TUCHKEVICH, YAVICH; LUTOVICH, KHARCHENKO). The diffusion and convection in a gas phase define more often the process rate through the regime, more preferable for the study of crystal growth mechanism, is that wherein the kinetics of surface process is a limiting one. The dependence of growth rate on surface characteristics (orientation, purity, etc.) is the criterion of kinetic region.

The analysis of gas-transportation process becomes complicated by the fact that usually a number of chemical reactions proceeds simultaneously in a system. The most wide-spread way of chosing of transportation conditions and the carrier remains still an empirical one in spite of a large number of works in transportation reactions. Usually the transportation reaction is postulated on the grounds of analogies with similar systems or according to the stability of the different hypothetical components of gas phase. The direction and rate of these reaction determine the efficiency of material transport process in a system. The physical-chemical investigation of separate reactions responsible for the material transport in a system and the totality of these reactions is interesting for finding out the chemism of gas transportation systems and for analytical description of transportation process.

3. The information not only of a gas phase composition and growth kinetics but also the information of a composition and structure of the absorption layer and crystallization surface is necessary to build up the crystal growth theory, which takes into account the heterogeneous reaction proceeding and the non-crystallizing atoms and molecules participation in surface processes.

2. Technique and computer experiment

The using of the Monte-Carlo method is suitable for the analysis of film growth processes by chemical reactions. The realization of theoretical models can be tested by the Monte-Carlo method. The dependencies of growth rate on any growth parameters: temperature, supersaturation, components relation in a gas phase, etc. can be found by the phenomenological consideration of the growth by the Monte-Carlo method. And the process limiting stage can be revealed according to these growth parameters. The process micromechanism which occur at a film growth by chemical transportation can be followed at microscopic approach. The Monte-Carlo method gives the possibility to test the realization of different reactions on a surface or in initial phase, to obtain the information of a gas phase composition, and of a surface structure and composition as well which contacts with a complicated gas phase under the conditions close to the equilibrium ones.

In present work the Monte-Carlo method is used for the growth conditions analysis of Si epitaxial films in gas transportation systems. The problem of obtaining of the growth characteristics and process micromechanisms, the problem of justification of the possibility of different reactions course and absorptive layer composition investigation have been defined. The authors consider that the investigation of Si growth conditions will help to produce the growth quantitative models of the other materials and also will help to estimate the effect of the new factors in the methods already used.

Two ways of Si obtaining in the chloride system have been analyzed by the computer simulation.

The possibility of the following reaction course (ALEKSANDROV et al. 1983) has been considered first:

$$SiCl_4 + 2\,H_2 \underset{v_2}{\overset{v_1}{\rightleftarrows}} Si + 4\,HCl \tag{1}$$

with the account that the surface etching can occur in hydrogen flow at high temperatures and high concentrations of $SiCl_4$:

$$SiCl_4 + Si \rightleftarrows 2\,SiCl_2. \tag{2}$$

The probabilities of each reaction course at a concrete point were determined through the equilibrium constants and were compared with random numbers. The equilibrium constants were calculated by Gibbs free energy change or by the known rates of v_1 direct and v_2 reverse reactions. The probability of reaction implement is assumed to be proportional to its rate, since $v_1 = K_1 v_2$. The probability of direct reaction course in one cycle or in time of $\Delta\tau = (v_1 + v_2)^{-1}$ has been determined as:

$$P_{1,\mathrm{d}} = v_1\,\Delta\tau = \frac{v_1}{K_1 v_2 + v_2} = K_1(1 + K_1)^{-1} \tag{3}$$

and for the reverse one:

$$P_{1,2} = (1 + K_1)^{-1} \tag{4}$$

The interaction of $SiCl_4$ molecule and two molecules of hydrogen molecules concentration is equal to $C_{H_2} = 1 - C_{SiCl_4}$, at a given concentration of $SiCl_4$ in hydrogen flow — C_{SiCl_4}. The probabilities of reagents collision (1) are assumed to be proportional to their concentrations of the first and second degree, respectively. If two following numbers occur to be less than these concentrations, the reaction (1) becomes possible. If the next random number occur to be less than P_{dir} (formula (3)), it is assumed that Si atom is formed and absorbed at a point on the surface, the surface coordinates are also found out by a random number. The surface height is changed at this place.

It is assumed while analyzing reaction (2), that $SiCl_4$ molecule can interact with a surface if $SiCl_4$ does not encounter two molecules of hydrogen. Thus, in the cases when $G < C_{SiCl_4}$ and $G < (1 - C_{SiCl_4})^2$ the conditions were not carried out, the probability of etching reaction course was analyzed from the relation:

$$P_{2,d} = K_2(K_2 + 1)^{-1} \tag{5}$$

where K_2 is the equilibrium constant of the reaction (2).

The supersaturation near a surface was determined via the sum of the reevaporations and isolation of atoms.

3. Results of simulation and their discussion

The kinetics of Si film growth has been investigated in accordance of the described model. The dependence of growth rate on time and temperature has been shown. The influence of surface steps on growing surface roughness (the microscopic point of view) during silicon film growth is considered.

The dependence of silicon film growth rate on the (111) plane on $SiCl_4$ concentration at hydrogen flow at $T = 1500$ K is shown in Figure 1. It is seen from the figure that the growth rate increased, reached the maximum and then decreased while the concentration increase. Thus, it can be assumed that (beginning from $C_{SiCl_4} = 6\%$) the surface gas etching of a growing crystal appears in the initial mixture according to the reaction (2). This etching retards the growth rate. The similar experimental dependences have been found out in the works of KHARCHENKO at silicon film production and GIVARGIZOV at Ge film production by the chloride method.

Thus, the reversibility of reaction (1) means that it can attribute both the growth and the etching, and the observed growth rate is the result of superposition of these two processes. At concentration increase of $SiCl_4$ the reaction (1) equilibrium is displaced directly, the etching role increases by this. The HCl concentration increase on the contrary increases the relative rate.

The change with time of roughess and Si growth rate on the (110) plane, containing a step is shown in Figure 2. The roughness maximum and the area with the smallest

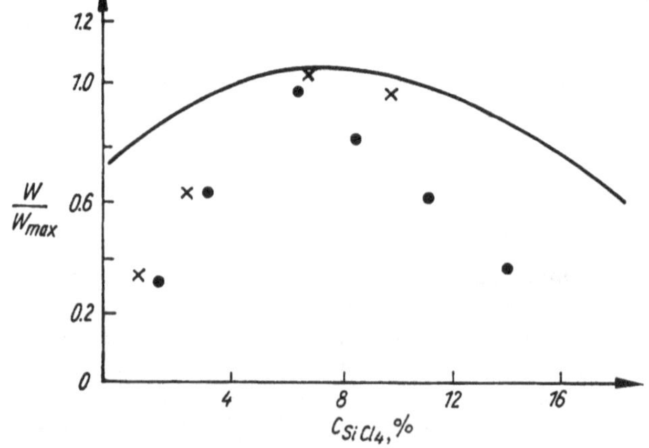

Fig. 1. The dependences of growth relative rate of Si films on $SiCl_4$ concentration in a gas phase ($T = 1500$ K) (× – the experiment of KHARCHENKO and ● – the experiment of GIVARGIZOV)

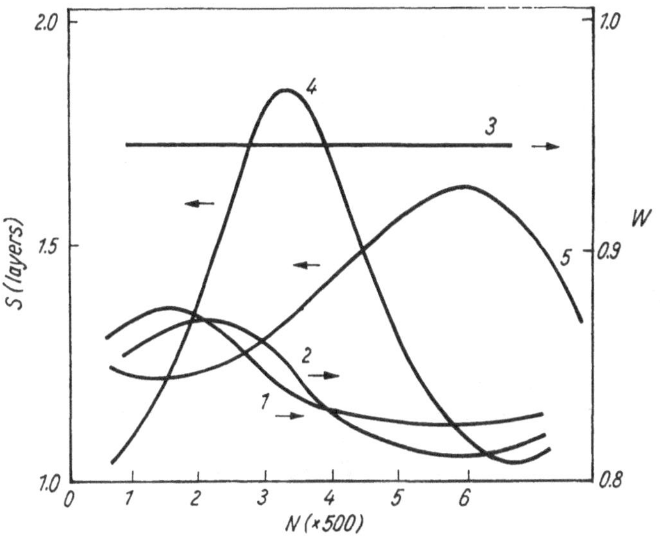

Fig. 2. The dependence of growth rate and (110) plane roughness of Si film, containing a step, on time duration (N is the number of computer cycles). The curves 1, 2 are the growth rates; the curves 4, 5 are the roughnesses, at 1300 and 1500 K, respectively, the curve 3 is the growth rate at the supersaturations more than 10 %

growth rate apparently corresponds to a new step formation: the adsorbed atoms readily return in a gas phase without rate increasing; the roughness in this case is large. The rate value does not go to zero, since the step echelon moves. It corresponds to the numerical values of roughness and it is confirmed by the computer calculations of a surface relief. The formation of stable step leads to the increasing of active places number for the crystallization and the growth rate increases. When the step passes the whole plane surface, the rate remains its value until a new stable step forms. Such type of dependence $W(N)$ occurs at the supersaturation up to 10%. At more high supersaturation the step personality disappears, the roughness increases, and the growth rate after 50 computer cycles (on the area site) becomes constant.

At the simulation of silicon films growth, which are obtained by the second way it is assumed, according to NISHIZAWA and NIHIRA, the course of some competing reactions to be possible, these reactions can be shown by the scheme:

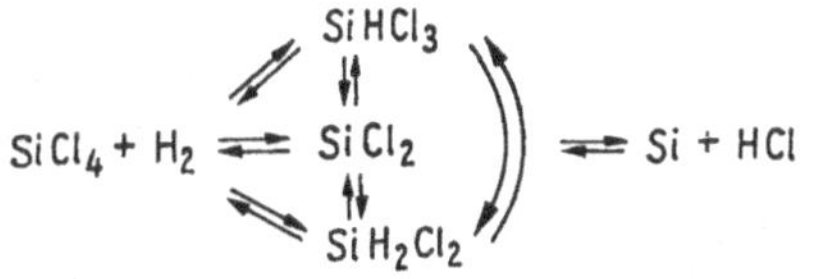

The interstitial particles $SiHCl_3$, SiH_2Cl_2, HCl, and $SiCl_2$, are formed as a result of $SiCl_4$ reaction with hydrogen. These particles while interacting with each other and with hydrogen lead to silicon adsorption on a surface.

The constant analysis of reaction equilibrium according to KUZNETSOV et al. has been shown that at the temperature interval 1000–1300 K the reaction (7) or (8) was possible:

$$SiCl_4 + H_2 \rightleftarrows SiCl_2 + 2\,HCl\,, \qquad (7)$$

$$SiCl_4 + H_2 \rightleftarrows SiHCl_3 + HCl\,. \qquad (8)$$

If reaction (8) has occurred as it was given in (10), $SiHCl_3$ was taken as an initial material, the reactions (9), (10) becomes possible undoubtedly:

$$SiHCl_3 + H_2 \rightleftarrows SiH_2Cl_2 + HCl\,, \tag{9}$$

$$SiHCl_3 \rightleftarrows SiCl_2 + HCl\,. \tag{10}$$

Silicon is obtained according to the reactions (11)—(13) by further decomposition of the products of the reactions (9) and (10).

$$SiH_2Cl_2 \overset{K_3}{\rightleftarrows} SiCl_2^* + H_2\,, \tag{11}$$

$$SiH_2Cl_2 \overset{K_4}{\rightleftarrows} Si^* + 2\,HCl\,, \tag{12}$$

$$SiCl_2^* + H_2 \overset{K_5}{\rightleftarrows} Si_{cr} + 2\,HCl\,. \tag{13}$$

The reaizaltion of any elementary event, as in the previous case, has been determined by the comparison of random numbers with corresponding transition probability. The atoms of silicon which belong either to the adsorption layer (Si^*) or to the crystal (Si_{cr}) are consireded to be on the silicon surface as well as the molecules of $SiCl_2^*$. In each case the events are analyzed in different way.

1. If the molecule SiH_2Cl_2 approaches to the site, where silicon of the crystal is placed the possibility of the reactions (12) or (11) course is tested by random number. If:

$$G < K_3(K_3 + K_4)^{-1} \tag{14}$$

the reaction (11) occurs, and the molecule $SiCl_2$ is adsorbed on a surface. If the condition (14) has not been carried out and the reaction (12) occured, Si^* is adsorbed on a surface, which has the possibility to build-in the surface with the probability:

$$L = \exp\left(\frac{2\varphi - E}{kT}\right)\left[\exp\left(\frac{2\varphi - E}{kT}\right) + \sigma^{-1}\right] \tag{15}$$

in the case if the random number $G < L$. In (15): φ is the energy of one bond, $E(l)$ is the energy of l bonds, $\sigma = N_{ad}/N_{ad}^{eq} = P_i/P_i^{eq}$ is the relative supersaturation which is determined either by the real and equilibrium partial pressures (P_i, P_i^{eq}) or by the number of atoms in real N_{ad} and equilibrium N_{ad}^{eq} adsorptive layers.

If the molecule HCl approached to the site containing Si_{cr}, the reverse reaction (13) occurs. The interaction of crystalline silicon with hydrogen is assumed to be impossible.

2. If the silicon of adlayer is at the surface site understudy, the analysis is carried out by the following way. The probability of the reverse reaction (12) is tested at the approach to the HCl molecule surface. In the case if the random number $G < (K_4 + 1)^{-1}$ the particle SiH_2Cl_2 is formed, on the contrary, Si^* which turns into Si_{cr} with the probability (15) appears on a surface. Si^* does not interact with molecules SiH_2Cl_2 and H_2 in the reactions (11)—(13).

3. If the molecule $SiCl_2^*$ is at the site either direct reaction (13) at $G < (K_3K_5 + 1)^{-1}$ or reverse reaction (11) is possible when the atom H_2 approaches the molecule $SiCl_2^*$. The attachment of the molecules SiH_2Cl_2 or HCl to a surface does not occur in this case. The description of chemism of the process of silicon production in the system $SiCl_4$ + HCl + H_2 has been done according to the given scheme. The description has been carried out as a complex of chemical reactions, which accompany the main one leading to the silcon precipitation and reaction products formation. Computer simulation allows to answer the main questions of any experiments of growth at chemical transport: where (in a gas phase or on a crystal surface) the silicon precipitates and how it crystallizes on a surface and what is the composition of adsorptive layer.

The reaction (12) gives the main contribution into a growth rate at all temperatures by the results of simulation. The contribution of this reaction increases with the temperature increasing. In this case Si* is formed, which then builds-in a crystal. The contribution of direct reactions (11) and (13) is practically equal ($\sim$ 0.03 from reaction (2)) at low temperature (1000 K). The contribution of these reactions decreases up to 1/6 from (2) at high temperatures. The Si_{cr} formation decreases more abruptly by the reaction (3). The content of Si_{cr} decreases with temperature at an adlayer, respectively, however, not so abrupt as at the reaction (3), since the quantity of Si* increases which then builds-in a crystal.

The reverse reactions (11)—(13) are realized slightly. At temperature increasing the realization of reverse reaction (11) with formation of SiH_2Cl_2 in a gas phase somewhat increases. The reverse reactions (12) and (13) at a given concentration of HCl ($C_{HCl} \sim 6 \cdot 10^{-3}$) practically do not occur.

Table 1
The adlayer composition of Si (III) surface in the system $SiCl_4$–HCl–H_2

Adsorbed particle	1000 K	1100 K	1200 K	1300 K	the components' concentration in a gas phase
Si_{cr}	0.71	0.80	0.82	0.78	without HCl
Si*	0.19	0.12	0.14	0.20	$C_{SiH_2Cl_2} = 0.3$
$SiCl_2$	0.10	0.08	0.04	0.02	$C_{H_2} = 0.7$
Si_{cr}	0.76	0.75	0.78	0.71	$C_{SiH_2Cl_2} = 0.3$
Si*	0.14	0.17	0.14	0.22	$C_{H_2} = 0.694$
$SiCl_2$	0.10	0.08	0.08	0.07	$C_{HCl} = 6 \cdot 10^{-3}$

The change of adsorptive layer composition with temperature is shown in Table 1. It is seen from the Table that the relation of components concentration at an adlayer is the following at all temperatures:

$$C_{Si_{cr}} > C^*_{Si} > C_{SiCl_2} .$$

With the Si_{cr} concentration increasing Si* decreases and alternatively. The concentration of HCl almost does not change. The crystalline silicon (Si_{cr}) covers of 70—80% of the surface, Si* — 13—20%, the particles $SiCl_2$ — 3—10%. The adlayer composition change is shown in Figure 3.

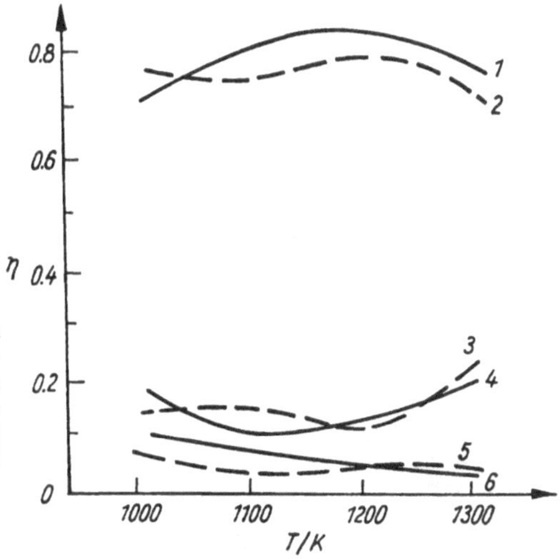

Fig. 3. The change of adsorptive layer composition in the system $SiCl_4$ + HCl + H_2 according to the temperature. The curves 1, 2 are the Si_{cr} atoms fraction in an adlayer; the curves 3, 4 are the Si* atoms fraction; the curves 5, 6 are $SiCl_2$ molecules fraction. Solid line — the growth in the absense of HCl, dotted line — the growth with HCl

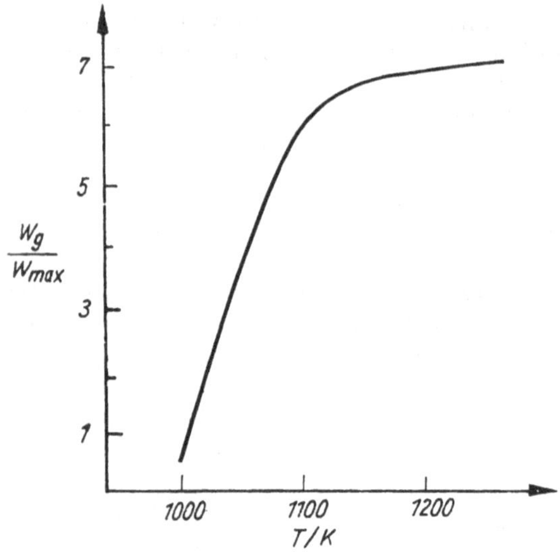

Fig. 4. The dependence of growth relative rate of Si films in the system $SiCl_4$ + HCl + H_2 on temperature

The existence of two areas has been found out while making investigation of the dependence of growth rate on temperature (Fig. 4): the growth rate depends on temperature strongly at low temperatures (to 1100 K) and the rate almost does not depend on temperature at that above 1100 K. Therefore, the growth rate is determined by the surface factors (kinetic regime) over I area, and by the material transport to a surface (diffusive area). Such dependence course corresponds to the experimental results (GIVARGIZOV; LUTOVICH, KHARCHENKO; STRELCHENKO, MATYASH), according to which linearly increases with temperature increasing over the low temperatures area, and the rate becomes constant at higher temperatures.

The dependence of growth rate on the main material concentration in a gas phase is shown in Figure 5. The growth rate increases with the concentration of SiH_2Cl_2 on the account of Si_{cr} contribution. The contribution of Si* and $SiCl_2$ into the total rate de-

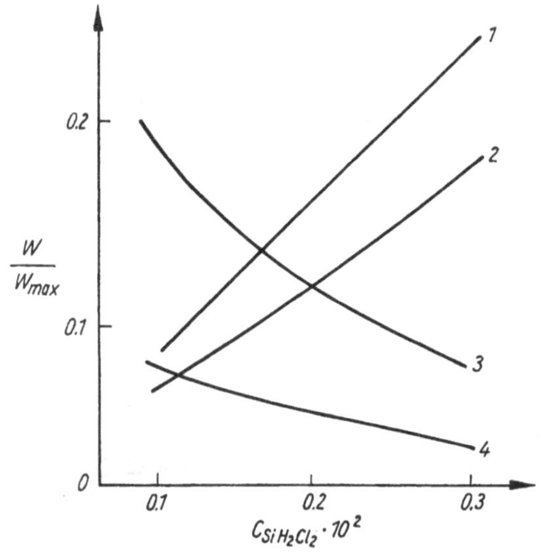

Fig. 5. The dependence of growth relative rate on main component concentration in a gas phase. The curve 1 is the total growth rate, the curves 2, 3, 4 are the contribution into the growth rate of Si_{cr} Si* and $SiCl_2$ atoms, respectively

creases with the concentration increase. The same shape of a curve has been observed in the experiment by ARIDZUMI.

4. Conclusion

The investigation by the Monte-Carlo method of the growth of the silicon epitaxial films in a chloride gas-transport system has allowed to find out the composition of adsorptive layer, the micromechanism of the reactions of Si atoms building-in into the growing crystalline layer and the growth conditions influence on the growth rate and film surface roughness.

The probabilities of course of etching and growth reactions and the intermediate compounds formation have been determined by the reaction equilibrium constants and initial materials concentrations and have been compared with random numbers. The main results are:

1. The concentration of $SiCl_4$ providing the maximum rate of film growth has been established.

2. The change of film growth and its surface roughness in time (with a step on a surface) depending on growth temperature and supersaturation has been shown.

3. The change of adsorptive layer composition in the system $SiCl_4$–HCl–H_2: fraction of adatoms, silicon atoms built-in a crystal and molecules $SiCl_2$, depending on temperature has been determined.

4. The change of silicon film growth rate depending on temperature and concentration change of SiH_2Cl_2 has been established and the contribution of growth mechanism (with participation of adatom, silicon atoms and molecules $SiCl_2$) into the total rate of film growth has been shown.

The regularities of film growth under the conditions of chemical transport have been established by the Monte-Carlo method to be in agreement with known experimental and estimated data (RUSAIKIN; ALEKSANDROV et al. 1986), these regularities also reveal the pecularities of the formation processes of silicon film structure unknown before.

References

ALEKSANDROV, L. N., KOGAN, A. N., BOCHKOVA, R. B., DIAKONOVA, V. I., TROSTINA, N. P.: in: Computer simulation of radiative defects in crystals (ed. YU. V. TRUSHIN), Leningrad 1983, p. 174

ALEKANDROV, L. N., KOGAN, A. N., BOCHKOVA, R. V., TROSTINA, N. P.: Kristallographiya **30** (1985) 236 (a)

ALEKSANDROV, L. N., KOGAN, A. N., DIAKONOVA, V. I., TROSTINA, N. P.: Poverkhnost: Fizika. Khimia. Mekhanika **3** (1985) 135 (b)

ALEKSANDROV, L. N., KOGAN, A. N., TIKHONOVA, N. P.: phys. stat. sol. (a) **92** (1985) 109 (c)

ALEKSANDROV, L. N., KOGAN, A. N., BOCHKOVA, R. V., DIAKONOVA, V. I., TIKHONOVA, N. P.: in: 7th Conference in processes of growth and synthesis of semiconductor crystals and films. Abstracts, v. 1, Novosibirsk 1986, p. 217

ARIDZUMI, T.: Kindzoku hemen gidzusy **32** (1981) 423

GIVARGIZOV, E. I.: Fiz. Tverd. Tela **5** (1962) 1150

KHARCHENKO, V. V.: The problems of Si epitaxial deposition, Tashkent 1976

KUZNETSOV, F. A., BUZHDAN, YA. M., FEDOROVA, G. V., KUZNETSOV, G. N.: in: The processes of growth and synthesis of semiconductor crystals and films, part 1 (eds. L. N. ALEKSANDROV, F. A. KUZNETSOV), Novosibirsk 1975, p. 69

LUTOVICH, A. S., KHARCHENKO, V. V.: in: Thin film crystallization (ed. V. A. ARIFOV), Tashkent 1970, p. 3
NISHIZAWA, J., NIHIRA, H.: J. Cryst. Growth **45** (1978) 82
RUSAIKIN, M. P.: In: The growth of semiconductor crystals and films, v. 2 (ed., F. A. KUZNETSOV), Novosibirsk 1984, p. 21
STRELCHENKO, S. S., MATYASH, A. A.: The review in electron technique ser. 6, Materials 8 (1984) 1054
TUCHKEVICH, V. V., YAVICH, B. S.: In: The growth processes of crystals and semiconductor films, (ed. L. N. ALEKSANDROV, L. A. BORISOVA), Novosibirsk 1970, p. 232

(Received July 4, 1987)

Author's address:
co Prof. Dr. L. N. ALEKSANDROV
630090 Novosibirsk, 90
pr. Akad. Lavrentyeva, 13
Institute of Semiconductor Physics
Siberian Branch of the Academy of Sciences of the USSR
USSR

Cryst. Res. Technol.	23	1988	2	153—158

F. M. Alemaykin, A. N. Kogan, T. B. Yudina

Mordovian N. P. Ogaryov State University, Saransk, USSR

Study of the Growth Mechanism in the ADP-type Anion-doped Crystals

A simultaneous simulation of the growth process in the A(K)DP-type anion-doped crystal and growing of this crystal from aqueous solution under controlled conditions has made it possible to analyse the mechanism of impurity influence on the growth rate of various crystal faces. The anisotropy of the face growth rate and the capture coefficient of various impurities by the faces have been determined by the difference in dissolution heats and in sizes of replacing ions. The increase in energy change, when the ion in a solution enters a crystal, causes an increase in the growth rate and the capture of impurities by the face with the largest change. Breaking of one of the joined faces and increased growth of another one resulted in forming of the new faces, i.e. in changing crystal morphology.

Одновременное моделирование процесса роста анионолегированного кристалла типа А(К) ДП и выращивание такого кристалла из водного раствора в контролируемых условаиях позволили выяснить механизм влияния примеси на скорость роста различных граней кристалла. Установлено, что анизотропия скорости роста граней и коэффициента захвата ими различных примесей определяются различием в теплотах растворения и размерами замещающих ионов. Повышение изменения энергии при вхождении иона из раствора в кристалл обусловливает увеличение скорости роста и захват примеси той гранью, для которой это изменение наибольшее. Торможение одной из стыкующихся граней и ускоренный рост другой приводят к образованию новых граней, то есть изменению морфологии кристаллов.

Ammonium dihydrogen phosphate (ADP)-type crystals produced by the method of temperature decrease and evaporation in a dynamic condition display differences in the morphology for large supersaturations or impurity presence in an aqueous solution. Anisotropy of the growth rate is also observed.

A computer simulation of the atom dynamics on the surface makes it possible to establish the influence of the experimental parameters, such as type and quantity of impurities, supersaturation, temperature, etc. upon the growth process, and thus, upon the crystal morphology, as a whole.

The present paper is aimed at establishing factors responsible for a change in the crystal form, as well as their mechanisms acting in different conditions of the growth with the application of real and computer experimental results.

1. Experimental results

The growth of ammonium dihydrogen phosphate (ADP) and potassium dihydrogen phosphate (KDP) crystals has been made at temperatures of 30—65 °C from pure solutions of $NH_4H_2PO_4$ and $K_2H_2PO_4$ in the presence of the impurities $(NH_4)_2CrO_4$, $(NH_4)_2SO_4$, $(NH_4)_2MoO_4$ (their concentration being changed from 1 to 30%).

The pure ADP and KDP crystals grew with a normal morphology — having a small coneness of 6—8°. The increase of supersaturation up to 18% and the crystallisation temperature to 65 °C did not essentially effect the coneness, if pH of solution ranged from 4.2 to 4.5.

The results of morphological analysis show different influences of the impurities. Crystals grew with a normal morphology at concentrations of $(NH_4)_2SO_4$ up to 10% and supersaturations from 2 to 10%. A further increase of impurities in the solution and of the supersaturation up to 18% resulted in the widened morphology of faces (SOLONENKO). The ADP crystals were produced after the addition of $(NH_4)_2SO_4$ up to 30%, and the KPD crystals up to 20%. The addition of $(NH_4)_2SO_4$ up to concentrations of 10% at a supersaturation of 1–2% in the solution has been found to effect a more isometric and perfect ADP crystal growth. At the introduction of K_2CrO_4 from 1 to 20%, the KDP crystal wedging out did not exceed 2–6° or was lacking all together. When $(NH_4)CrO_4$ increased to 35%, the ADP crystals had a dark-green colour and distinct conical shape.

The introduction of $(NH_4)MoO_4$ in an $NH_4H_2PO_4$ solution from 8–15% resulted in crystal top transformation and new faces developed forming a dyadical top with an angle of 119 ∓ 2°.

A further increase in the impurities resulted in a complete change of the ADP crystal surface form. It has been established (ALEMAYKIN, SOLONENKO) that under investigated conditions the crystal growth occurs in an kinetic mode. The growth rates were measured by the optical method, their supersaturation dependence is shown in Figure 1. A comparison of curves 1, 3 shows that the (110) face grows faster than the (100) face for all supersaturations. The (100) face growth rate decreases at 12% of $(NH_4)_2CrO_4$ (curves 1, 5); under high supersaturations in the presence of $(NH_4)_2CrO_4$ the (110) face grows faster than from pure solution (curves 3, 7).

The growth rate of a pure (100) face increases linearly with supersaturation. In all other cases the supersaturation growth rate dependence is more complicated. The results of the analysis of the growth rate temperature dependence of the pure (110) face $v = v_0 \exp(-Q/kT)$ and with impurities (curve 1, 3) as $\ln v/v_0 = f(1/T)$ at $\sigma = 3\%$ are presented in Figure 2. The points lie on the line well, that shows the common growth process. The presence of chromium increases the growth activation energy Q

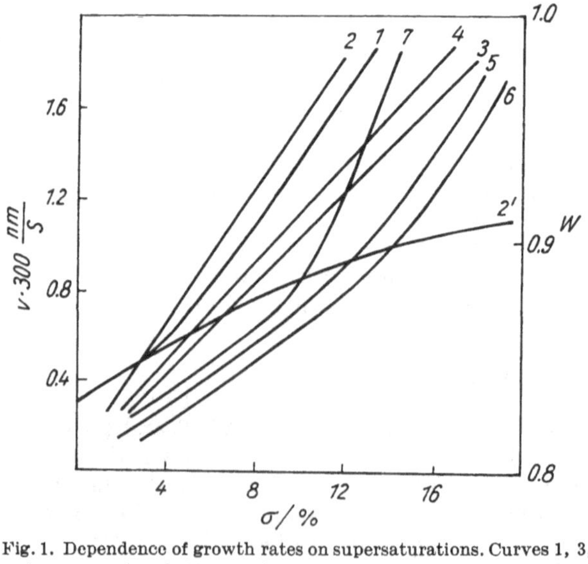

Fig. 1. Dependence of growth rates on supersaturations. Curves 1, 3, 5, 7 – experimental; 2, 2', 4, 6 – computer, 2, 2', 1, 5, 6 – (100), 3, 4, 7 – (110)

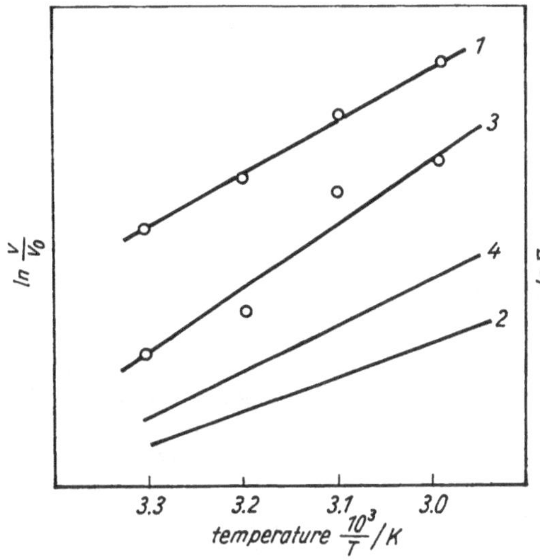

Fig. 2. Temperature dependence of growth rates (110) for $\sigma = 3\%$. Curves 1, 3 — experimental; 2, 4 — computer

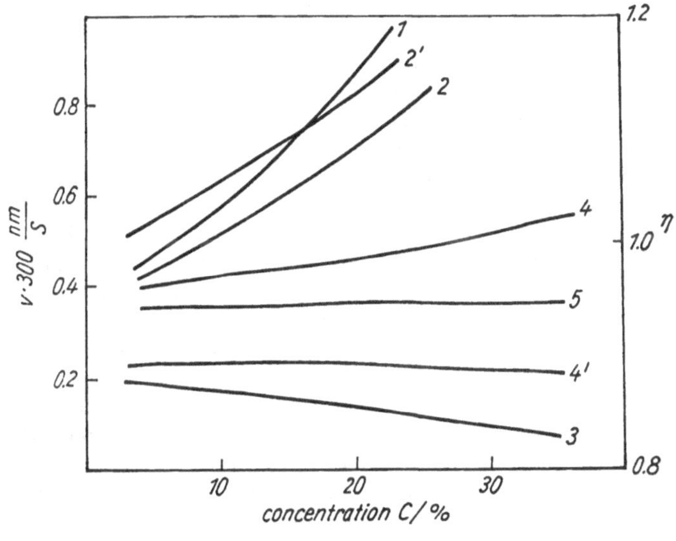

Fig. 3. Dependence of experimental growth rates (curves 1, 3, 5) in presence of $(NH_4)_2MoO_4$ and calculated growth rates with $\Delta\varphi_2 = 1.3\ \Delta\varphi_1$ (curves 2, 4) and with $\Delta\varphi_2 = 0.8\ \Delta\varphi_1$ (curves 2′, 4′), coefficient of impurity capture η, curves 1, 2′, 2 — (110) face; 3, 4′, 4 — (100); 5 — (111) face

by 7 kJ/mol. For the (100) face, the increase of Q makes 28 kJ/mol for 12% of $(NH_4)_2 \cdot \cdot CrO_4$ in solution. Concentration dependences of growth rates in ADP crystal of various faces in the presence of $(NH_4)_2MoO_4$ are given in Figure 3 (curves 1, 3, 5). A different character of the impurity effect on various faces is well seen. The (110) growth rate increases with the concentration in C, that results in its degeneracy. The growth rate of the faces (100) and (111) decreases with an increases in C.

2. Computer experiment and its results

Structural units in the computer crystal model were anion and cation NH_4^+, $H_2PO_4^-$, SO_4^{--}, $CrOO_4^-$ — type complexes that form a tetragonal lattice. An elementary cell of this lattice consists of the 4 planes with various arrangements of the complexes (Alemaykin et al.). Each ion has two nearest neighbours at a distance of r_0, the next 4 being at a distance of r_0/a_1 and 4 at a distance of r_0/a_2. These distances are termed to be the length of binding. For pure ADP $a_1 = 0.86$, $a_2 = 0.7$. During the process of the growth simulation the anions were prohibited to occupy cation places (CHERNOV, LEWIS). For one analysis of the probability of complex attaching the face that possesses n surface sites, as many as 1 to 3 diffusion exchanges have been simulated between an ion and its nearest empty site of the same type depending on the temperature growth. In other respects, the simulation technique is similar to that one described by ALEKSANDROV et al.

During the transition of the complex from a solution to the crystal energy change $\Delta\varphi$ calculated for one binding with the account for its length was valued by means of the dissolution heats of the main components and impurities ΔH_i.

$$2\Delta H_i = 2\Delta\varphi_i + 4a_1\Delta\varphi_i + 4a_2\Delta\varphi_i . \quad (1)$$

The difference in the anion sizes at the substitution of main anions by impurities caused the deformation energy which was also recalculated for one binding. The sum of these energies was included in the calculation of transition probabilities. The values of such simulation parameters as the equilibrium dissolutions of the components in solutions containing alloy additions were determined experimentally. The effective supersaturations for the growing face that were recalculated from the experimental ones with the account for the crystal rotation rate were introduced into a computer model. Display printing was made for: the dimensionless growth rates W, that characterizes a part of atoms attached to a crystal in a total amount of atoms, arriving onto the surface; the impurities capture coefficient η, the number of realized diffusion exchanges Γ for each sort of complexes on the (100) and (110) faces calculated for one surface pack and roughness S determined as a dispersion over the height. We took into consideration the interaction in three coordination spheres (each ion has 10 neighbours). A limitation by two coordination spheres (each ion has 6 neighbours) resulted in a change in analysed parameter values by 20% with no change in the common behaviour of the curves. Figure 4 shows a fragment of the dependences of the growth rate W and the roughness S expressed in growth layers upon the number of computer cycles N at $T = 40$ °C. For $\sigma = 5$ at an increase in calculation duration, the picture is regularly repeated, the difference being not more than 10% and the error being less than 5%. The maximum of roughness and the area with the lowest growth rate correspond to the formation of a new step: adsorbed atoms easily returning into a solution producing no increase in the growth rate, the roughness being essential. The value of the rate does not drop to 0 as it has been reported by MALKIN, because of the motion of the step train. This corresponds well to the numerical values of roughness and is confirmed by computer displays of the surface relief. The formation of a stable step results in an increase in the number of active places for attaching ions and the growth rate increases. When this step passes all over the face surface, the growth rate decreases and its value remains until a new stable step has been formed. This kind of dependence $W(N)$ is valid for supersaturations up to 10%. For higher supersaturations the individuality of the steps disappears, the roughness increases and the growth rate becomes constant after 50 computer cycles per one site of a simulation area (curve 3). Regarding the process of step formation revealed by simulation as the activation process, from the temperature dependence of duration of the area with the minimum rate $N \sim \exp\left(\frac{u_{ef}}{k \cdot T}\right)$. One may estimate an effective activation energy of steps formation — u_{ef}. For the (110) face of pure ADP, in recalculation per an ion, the value u_{ef} is somewhat larger than 1/3 of the dissolution heat, for the (100) face it is larger. In the presence of impurities with a larger $\Delta\varphi$ the step formation is easier.

Average values of the growth rates determined after 100 cycles per site are presented in Figure 1 for different supersaturations (even curves). For a more convenient com-

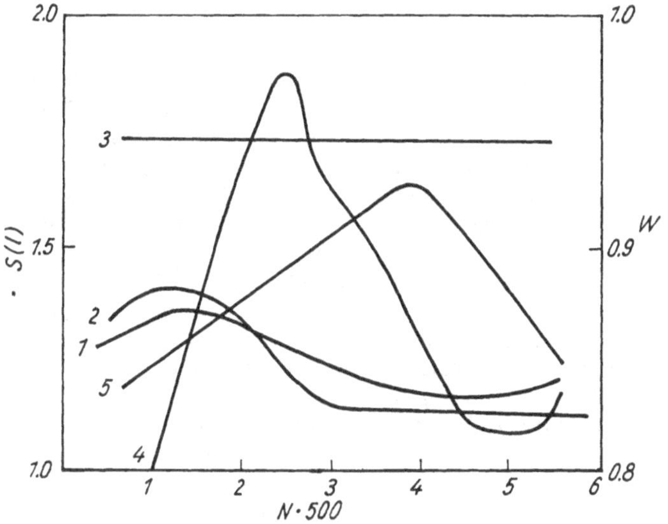

Fig. 4. Time dependence of the growth rate W (curves 1, 2, 3) and roughness S (curves 4, 5); 1, 5 – (110), $\sigma = 5$; 2, 4 – (100), 3 – (110), $\sigma = 20$

parison of experimental results the dimensionless computer rates W of the type of the $2'$ curve with a parabolic supersaturation dependence have been recalculated into the dimensional rates V by the following way. It has been supposed that at one point where $\sigma_1 = 3$, the dimensionless rate W_1 exactly corresponds to the experimental V_1. For other supersaturations the values V are found from the correlation which is already known ALEKSANDROV et al.)

$$V = V_1 \cdot \frac{W}{W_1} \cdot \frac{\sigma + 1}{\sigma_1 + 1}.$$

The results of V-recounts coincide well with the corresponding experimental curves for pure substances (curves 2, 4, and 1, 3), as well as for impure ones (curves 6 and 5).

The dissolution heat of $(NH_4)_2CrO_4$is higher than that of $NH_4H_2PO_4$. Thus, it was accepted in the computer experiment that with the replacing of ions of $H_2PO_4^-$ by CrO_4^{--},the binding energy change at the crystallization $\Delta\varphi_2$ is $1.2 \cdot \Delta\varphi_1$. With no account made for the deformation which is caused by the difference in the ion radii of the replacing ions by 30% and which is essential in the first and the second coordination spheres, such impurities must speed up the growth of both faces. The presence of the deformational energy reduces this effect, especially for the (100) face. In general, this face grows slower, because in the adsorption position the atom binding energy on the (100) face is less than that on the (110) face, and the deformation in the presence of the impurities becomes greater. The simulation of the presence of atoms attaching and diffusion on the (100) and (110) faces in the presence of impurity with $\Delta\varphi_2 = 1.3 \cdot \Delta\varphi_1$ has shown that for $\sigma = 20\%$ and at $T = 40$ °C after 200 cycles per site one can see the difference in the height of the faces which is larger by 2 growth layers and the formation area with atoms arrangements is as on the (111) face. With an increase in supersaturations and concentrations of the impurities this effect may be of a macroscopic character which has been experimentally observed.

The impurity with a more less dissolution heat of the $(NH_4)_2SO_4$ type for which $\Delta\varphi_2 = 0.8 \cdot \Delta\varphi_1$ practically does not change the growth rate at $\sigma = 20\%$ and but at all temperatures essentially decreases the roughness of the (110) face. Probably due to this, at small concentrations $(NH_4)_2SO_4$, there takes place a more perfect ADP crystals growth.

The simulation of diffusion processes on the surface with a preferability of atoms jumps in the direction of crystal rotation has shown that the number of realized exchanges increases with a temperature. The atoms with a larger $\Delta\varphi$ make a larger contribution. Figure 2 shows the temperature dependence $\ln \Gamma$ for an equal member of ions coming from the solution on a site of the (110) face and for a supersaturation of 5%, curve 2 being for the pure ADP, and curve 4 being for 12% of impurity with $\Delta\varphi_2 = 1.3 \cdot \Delta\varphi_1$. In the second case, the activation energy is higher by 7%. In both cases, it is less than the experimental energy of the growth rate activation which is determined by curves 1 and 2. The difference in values of these energies is approximately equal to the energy of step formation. In the simulation process, the coefficient of impurity captures η was also determined. Figures 3 shows its values for different faces and $\Delta\varphi_2$. For the (110) face $\eta > 1$, regardless of $\Delta\varphi$. The (100) face captures impurities badly, especially if $\Delta\varphi_2 < \Delta\varphi_1$ (curves 4' and 4). In both cases, its coefficient of capture is less than 1. At the same time, the diffusion exchanges intensity was analysed. It is large for $\Delta\varphi_2 > \Delta\varphi_1$. Probably, the increase in the (110) face growth rate with an increase in impurities cncentration in the solution may be largely caused by the diffusional processes. The impurity with lower $\Delta\varphi$ is less diffusion active. It is evident from a large number of repeated ion estrangements which have an impurity as the neighbour. These values are displayed as any of simulation results.

3. Conclusions

The simulation of ions dynamics on the growing from the aqueous solution ADP crystal face has shown the mechanism of the growth rate slowing down by impurities with a smaller dissolution heat. The impurities easily attach to steps as well as from them, thus prventing the attaching of main substance anions. Impurities with a larger $\Delta\varphi$ diffuse faster across the surface and adhere to the steps not preventing the attaching of main substance ions with nearly equal radii.

A comparison of the morphological analysis data with the results of the simulation of joined faces growth shows that the formation of new faces may be caused by the stimulation of a rapid growth of one of the edges.

References

ALEMAYKIN, F. M., KOGAN, A. N., UDINA, T. B.: 4th All-Union Symposium on Process of Crystal Growth, Heads of Reports **11** (1985) 16

ALEMAYKIN, F. M., SOLONENKO, T. B.: 4th All-Union Symposium on Process of Crystal Growth, Sakhkadzor 1972

CHERNOV, A. A., LEWIS, J.: Phys. Chem. Solids **28** (1967) 2185

ALEKSANDROV, L. N., KOGAN, A. N., BOCHKOVA, R. V., TROSTINA, N. P.: Kristallografiya **30** (1985) 236

MALKIN, A. I.: 4th All-Union Symp. on Process of Growth of Crystals. Sakhkadzor **2**, (1985) 58

SOLONENKO, T. B.: Materials of Young Scientists Symposium of Volgo-Vjatka region, Yoshkar-Ola 1973

(Received 1) February 8, 1987,
2) May 29, 1987)

Authors' address:

F. M. ALEMAYKIN, A. N. KOGAN, T. B. YUDINA
Mordovskaya ASSR
430 090 Saransk, USSR
Mordovian N.P. Ogaryov State University

Cryst. Res. Technol.	23	1988	2	159–164

A. M. Radwan

Theoretical Physics Laboratory, National Research Centre, Cairo, Egypt

A Study of the Effect of Depletion Charge on the Stable Crystals Structures of Heavy Alkalis

In the framework of second-order perturbation theory and model potential formalism we investigate the effect of variation of depletion charge on the prediction of stable crystal structures of K, Rb, and Cs. In the formal calculation the HCP structure is found to have the lowest energy for all three metals, even though the phonon contributions were added in case of K. Nevertheless, we were able to predict correctly the observed BCC structure by changing the sign of the depletion hole charge.

Im Rahmen der Störungstheorie zweiter Ordnung und des Modellpotentialformalismus wurde der Einfluß der Änderung der Verschiebungsladung auf die Bestimmung von stabilen Kristallstrukturen von K, Rb and Cs untersucht. In der formellen Berechnung hat die hdp-Struktur die niedrigste Strukturenergie für alle drei Metalle, selbst wenn im Falle von K die Phononenbeiträge dazu addiert wurden. Dennoch, eine richtige Bestimmung der experimentell beobachteten Krz-Struktur wurde durch den Wechsel des Vorzeichens der Verschiebungsladung erzielt.

1. Introduction

Theoretical prediction of the stable crystal structures of the heavy alkalis K, Rb and Cs was expressed by some authors (Schneider, Stoll; Mc Laren, Sholl; Moriarty; So, Wang; Upadhyaya, Wang) using different forms of model and pseudopotentials. All of them failed to predict the observed BCC structure of those three metals, except Schneider and Stoll for Rb and Cs (but not for K) and Upadhyaya and Wang for K and Cs (not for Rb). Hafner investigated K and Rb and obtained successfully their stable crystal structures. He has tried to improve the method of pseudopotential from first principles. However, in trying to match his calculated results with experiment he has actually parametrized the pseudopotential from first principles by replacing the theoretically calculated coefficients describing shell- and conduction exchange by parameters empirically chosen on the principle of best agreement. Although this did improve the agreement between theoretical and experimental results, in our view the introduction of empirical parameters is unjustified and only worsens the scheme proposed by Harrison. Stable structure of K alone was examined in a number of papers (Pick; Shaw 1969; Hafner, Nowotny; Paasch et al.; Williams, Appapillai; Smirnov et al.). In most of them the HCP structure was obtained instead of the observed BCC of K, except by Hafner and Nowotny, and by Smirnov et al. The above mentioned comment on the results of Hafner still holds in case of Hafner, and Nowotny. As for Smirnov et al. they used in their calculations for K an equilibrium volume 8% larger than the experimental atomic volume V_0 at which our calculations were performed. Now the question rises, which stable structure of K would they obtain using V_0.

In the present paper we discuss the different factors that may contribute to a correct prediction of the stable crystal structures of K, Rb, and Cs.

2. Scheme of Calculation

The structural energy expressed as a dimensionless coefficient is given by (Harrison):

$$\alpha_{st} = \alpha_{es} + \alpha_{bs} \tag{1}$$

α_{es} is the electrostatic coefficient and α_{bs} is the band structure contribution:

$$\alpha_{bs} = 2r_0/Z^{*2} \sum_{q \neq 0} |S(q)|^2 \, F(q) \tag{2}$$

r_0 is the atomic radius, Z^* the effective valence, $S(q)$ is the structure factor and $F(q)$ the energy-wave number characteristics. α_{bs} is usually calculated by interpolating the tabulated $F(q)$ to obtain its value at the reciprocal lattice vectors $\boldsymbol{g}$, since $S(\boldsymbol{g}) \neq 0$ and $S(q) = 0$ otherwise.

The usual procedure in case of simple metals is to calculate α_{st} of the three common crystal structures FCC, BCC and HCP at a fixed volume, usually, the experimental one. The crystal structure with the lowest energy α_{st} will be, theoretically seen, the most stable one. The differences $\Delta\alpha_{st}$ between the different crystal structures are usually very small. Thus to avoid any possible interpolation errors we calculated $F(q)$ directly at the reciprocal lattice vectors $\boldsymbol{g}$. Here a great deal of care should be given to the questions of numerical integrations at points close to the singularity at $q = 2k_f$ and to the convergence of α_{bs} (see Radwan, Saleh).

Now, from the electronic structure of the heavy alkalis it is noted that the d-bands lie just above the Fermi level (Cohen, Heine). Heine, Weaire; and Williams, Appapillai considered this as the most probable cause of the incorrect prediction of the stable crystal structure of K. However, Moriarty (1974) has developed the generalized pseudopotential theory GPT using a zero-order pseudoatom approach and has calculated α_{st} of K, Rb and Cs with and without s-d hybridization. In both cases he found HCP as the structure of lowest energy. His results show that the effect of hybridization is small and its inclusion does not alter the situation. Therefore, we did not consider the s-d hybridization in our calculations.

Williams examined the importance of third order contributions to α_{st} using a local model potential. He found that K and Rb have very small third order effects. McLaren, Sholl, and Rahman et al. calculated third order terms which are, respectively, 40% and 34% of the second order band structure energies α_{bs} of K, Rb, and Cs. However, the stable structures predicted from second order theory are the same as those obtained from third order for all three metals, namely HCP (McLaren, Sholl) and BCC (Rahman et al.). Thus, third order contributions were not taken into consideration, because they play a quantitative role and not a qualitative one that might influence the prediction of stable crystal structures of K, Rb, and Cs.

Moriarty (1974) suspects that the observed BCC structure in the heavy alkalis may be explained by considering the phonon contribution, i.e. the zero point energy, to the total free energy of the metal. This was also confirmed by McLaren and Sholl, since the quantity usually calculated is the free energy for a static lattice whereas physically it is the total free energy that must be a minimum. Thus we have to include the ground state energy of the phonons in our calculations.

The importance of the effective valence Z^* was pointed out by Radwan and Saleh. It was suggested (Radwan, Saleh; Radwan, Ali) that Z^* could be used as a parameter to obtain theoretical results having better agreement with the experimental ones. In fact, Z^* can be written as (Appapillai, Williams)

$$Z^* = Z - d. \tag{3}$$

Z is the chemical valence and d is the charge of the depletion hole, analogous to the orthogonalization hole of pseudopotential theory. The depletion hole represents the difference between model (smooth) and true (oscillating) wave functions due to the oscillations of the latter in core region. SHAW (1968) has approximated the depletion hole by a point charge (δ-function), whereas APPAPILLAI and WILLIAMS have taken the depletion charge to be uniformly distributed within a sphere of an averaged radius. This leads to a q-dependent d (spherical Bessel function) instead of the constant d of SHAW. In the present calculation we allow Z^* to change between $Z^* = Z - 2d$ and $Z^* = Z + d$ to investigate the effect of its variation on the prediction of stable structures of K, Rb, and Cs.

3. Results and conclusions

We have used the optimized model potential (SHAW 1968) and second order perturbation theory. The detailed equations are given in (RADWAN, SALEH). The exchange and correlation function is that of SINGWI et al. The band structure energy α_{bs} is calculated exactly from $F(q)$ at the reciprocal lattice vectors $\boldsymbol{g}$ without using interpolation at all. The model potential parameters used are those of APPAPILLAI and WILLIAMS (AW), and ESE and REISSLAND (ER) in case of K only. We have taken K as a representative of the heavy alkalis to study the following problems:

(1) Numerical integration at $\boldsymbol{g}$ close to the logarithmic singularity at $q = 2k_f$: this was solved by taking more Gauss points, namely, $N = 32$ for $\boldsymbol{g}$ close to $2k_f$ and $N = 10$ otherwise.

(2) Convergence of α_{bs} in K-space: this was examined carefully up to the extremely long range of $q = 17k_f$. It was found advisable to use $N = 12$ Gauss points for $q > 14k_f$. The convergence of α_{st} of K using AW potential and $Z^* = Z - d$ is shown in Figure 1. The most stable structure is the HCP. Although the summation was carried out to $q > 17k_f$, it suffices for qualitative cmparison to sum up to $q = 14k_f$ only.

(3) Effective mass approximation: this was included in our calculations according to the equations given by APPAPILLAI and WILLIAMS. The results for potassium are given in Table 1 using AW potential. AW* and AW** denote results obtained using effective masses $m_k/m_E = 0.6475$ and 1.1466, respectively.

From Table 1 we see that inclusion of effective masses does not change the sequence HCP — FCC — BCC of stable structures predicted before.

Table 1
α_{st} of Potassium calculated with (AW*, AW**) and without (AW) effective mass approximation

	BCC	FCC	HCP
AW*	−1.793657	−1.793872	−1.793897
AW	−1.793834	−1.794124	−1.794173
AW**	−1.794493	−1.794864	−1.794933

Table 2
Contributions to α_{st} of Potassium

	BCC	FCC	HCP
α_{es}	−1.7918585	−1.7917472	−1.7916762
α_{bs}	−0.0027296	−0.0030737	−0.0032082
α^0_{ph}	0.0026685	0.0027121	0.0027176
α_{st}	−1.7919196	−1.7921088	−1.7921668

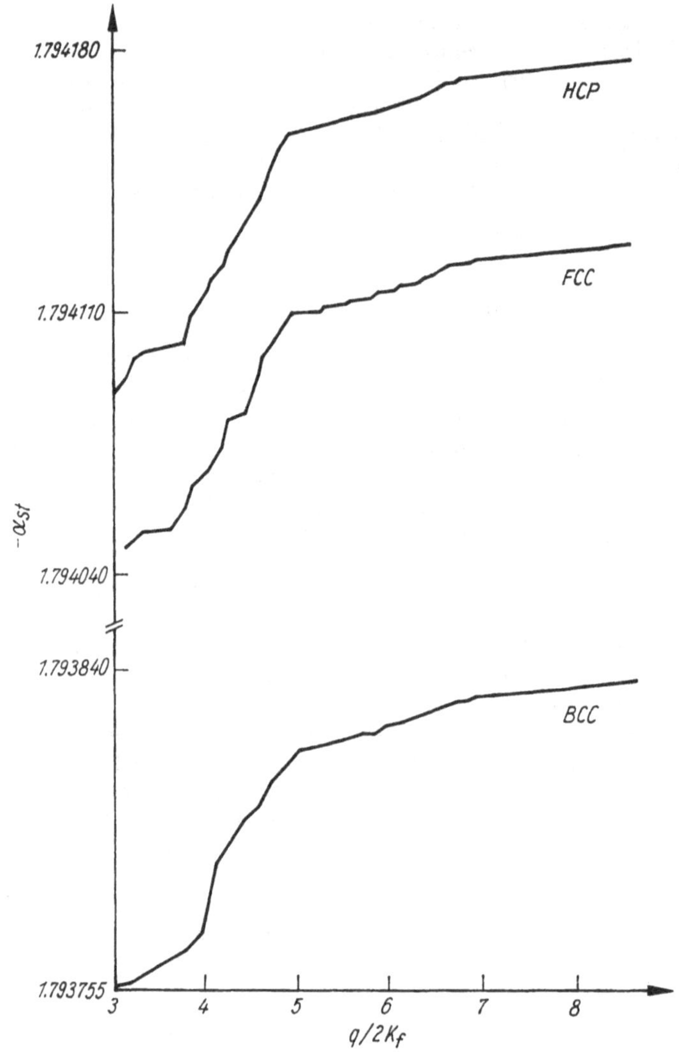

Fig. 1. Convergence of α_{st} of K. $Z^* = Z - d$ (AW potential)

(4) Phonon contributions: a term α_{ph} representing the contributions arising from the ionic motion is added to eq. (1). Since our calculations are performed at temperature $T = 0$ K, we need only the zero-point phonon energy α_{ph}^0 which was computed by PYNN and EBBSJÖ for K using the optimized model potential of SHAW (1968). Table 2 shows the different contributions to α_{st} of potassium, namely $\alpha_{st} = \alpha_{es} + \alpha_{bs} + \alpha_{ph}^0$, where α_{ph}^0 is taken from PYNN and EBBSJÖ, and we calculated α_{bs} using the same potential used by them.

From Table 2 it is noted that both α_{es} and α_{ph}^0 favour the BCC structure following the sequence BCC — FCC — HCP, whereas α_{bs} follows the sequence the other way round. However, the differences in α_{bs} are greater by an order of magnitude than those in α_{ph}^0. Therefore, the predicted most stable structure is again HCP instead of the observed BCC. Thus, inclusion of the lattice vibrations in the present form reduces the differences in α_{st} in favour of the observed BCC, but it does not remove the discrepancy between theory and experiment. Unfortunately, no such calculations of α_{ph}^0 of Cs and Rb are available.

(5) Variation of effective valence: calculations of α_{st} of K were carried out using ER potential and the following values of Z^*

a) $Z^* = Z - d = 1.12783$, this is the value of the formal theory,
b) $Z^* = Z - 2d = 1.25566$,
c) $Z^* = Z - 1/2d = 1.06392$,
d) $Z^* = Z - 1/4d = 1.03196$,
e) $Z^* = Z = 1$, i.e. depletion hole correction is not taken into consideration, and finally
f) $Z^* = Z + d = 0.87217$.

The results are given in Table 3.

Table 3
α_{st} of Potassium (ER potential) calculated with different values of Z^*

Z^*	BCC	FCC	HCP
$Z - d$ (Theory)	−1.793490	−1.793850	−1.793918
$Z - 2d$	−1.799325	−1.799876	−1.799991
$Z - \frac{1}{2}d$	−1.793940	−1.794194	−1.794237
$Z - \frac{1}{4}d$	−1.795407	−1.795605	−1.795637
$Z(d = 0)$	−1.797915	−1.798055	−1.798075
$Z + d$	−1.823038	−1.822922	−1.822903

Table 4
α_{st} of K, Rb and Cs (AW potential) calculated using different values of Z^*: a) $Z - d$, b) Z and c) $Z + d$

Element	Z^*	BCC	FCC	HCP
K	a) 1.14407	−1.793834	−1.794124	−1.794173
	b) 1	−1.800674	−1.800724	−1.800729
	c) 0.85593	−1.836109	−1.835882	−1.835862
Rb	a) 1.1525	−1.794308	−1.794675	−1.794748
	b) 1	−1.799136	−1.799223	−1.799239
	c) 0.8475	−1.836428	−1.836176	−1.836147
Cs	a) 1.1953	−1.795931	−1.796354	−1.796440
	b) 1	−1.800939	−1.800992	−1.801006
	c) 0.8047	−1.867376	−1.866949	−1.866916

From Table 3 we see that the structure of lowest energy follows the sequence HCP − FCC − BCC in all cases, except for $Z^* = Z + d$ where the order of stable structures becomes BCC − FCC − HCP. This means that we were able to predict the correct BCC structure of K by using $Z^* = Z + d$ instead of $Z - d$ as imposed by the theory. This success encourages us to apply the same procedure to Rb and Cs. The results are shown in Table 4 for K, Rb, and Cs using AW potential and three values of Z^*:

a) $Z - d$ of the formal theory,
b) Z, neglecting the depletion hole correction and
c) $Z + d$.

It is obious from Table 4, that the observed BCC structure is correctly predicted for K, Rb, and Cs by taking $Z^* = Z + d$, otherwise HCP is the structure of lowest energy. This means that, for the first time, the present method of optimized model potential and second order perturbation theory is capable of predicting the observed BCC structure of the heavy alkalis, provided that $Z^* = Z + d$ instead of $Z - d$ as implied by the formal treatment. In a recent paper RAHMAN et al. investigated the structural energetics of the heavy alkalis using third order perturbation theory and the non-local transition metal pseudopotential. They were successful in predicting the BCC structure for K, Rb, and Cs. We agree with their conclusion that even though the third-order contributions do not play a positive role in predicting the crystal structure of K, Rb, and Cs, they are important quantitatively in reproducing the correct Helmholtz free energies for these metals. Also the effects due to the depletion or orthogonalisation hole should be taken into account.

References

APPAPILLAI, M., WILLIAMS, A. R.: J. Phys. F**3** (1973) 759

COHEN, M. L., HEINE, V.: Solid State Physics **24**, New York 1970

ESE, O., REISSLAND, J. A.: J. Phys. F**3** (1973) 2066

HAFNER, J.: phys. stat. sol. (b) **57** (1973) 101

HAFNER, J., NOWOTNY, H.: Phys. Letters A **37** (1971) 335, and phys. stat. sol. (b) **51** (1972) 107

HARRISON, W. A.: Pseudopotentials in the Theory of Metals, New York 1966

HEINE, V., WEAIRE, D.: Solid State Physics **24**, New York 1970

McLAREN, R. E., SHOLL, C. A.: J. Phys. F**4** (1974) 2172

MORIARTY, J. A.: Phys. Rev. B **10** (1974) 3075

MORIARTY, J. A.: Phys. Rev. B **26** (1982) 1754

PAASCH, G., FELBER, B., ESCHRIG, H.: phys. stat. sol. (b) **54** (1972) K51

PICK, R.: J. Physique **28** (1967) 539

PYNN, R., EBBSJÖ, I.: J. Phys. F**1** (1971) 744

RADWAN, A. M., SALEH, A. A.: "Structural Energies of Light Metals", to be published in FIZIKA **19** (1987) 1

RADWAN, A. M., ALI, M. G. S.: Model Potential Calculation of Hexagonal Beryllium, to be published in Acta Physica Polonica, November 1987

RAHMAN, M. M., RAHMAN, S. M. M., CHOWDHURY, S. M. M. R.: J. Phys. F**14** (1984) 833

SCHNEIDER, T., STOLL, E.: Phys. Letters A **24** (1967) 258

SHAW, R. W.: Phys. Rev. **174** (1968) 769

SHAW, R. W.: J. Phys. C **2** (1969) 2335

SINGWI, K. S., SJÖLANDER, A., TOSI, M. P., LAND, R. H.: Phys. Rev. B **1** (1970) 1044

SMIRNOV, V. A., PODKORYTOV, S. I., KISELEV, YU. V., KOZOLOV, E. V.: Phys. Met. Metall. **47** (1980) 28

So, C. B., WANG, S.: J. Phys. F **7** (1977) 35

UPADHYAYA, J. C., WANG, S.: J. Phys. F **10** (1980) 441

WILLIAMS, A. R.: J. Phys. F **3** (1973) 781

WILLIAMS, A. R., APPAPILLAI, M.: J. Phys. F **3** (1973) 772

(Received June 24, 1987)

Authors address:

Dr. Ahmed M. RADWAN
National Research Centre, Theoretical Physcis Laboratory
Tahrir Street
Cairo-Dokki
Egypt

Cryst. Res. Technol. | 23 | 1988 | 2 | 165–169

I. S. Vassilev, L. D. Pramatarova, M. B. Baeva

Institute of Solid State Physics, Bulgarian Academy of Sciences, Sofia, Bulgaria

X-ray Topographic Study of GaSb Substrates for Epitaxial Layer Growth

The depth of the layer damaged by silicing of GaSb ingots has been determined by X-ray topography. The changes in the defect structure of substrates after successive etching to different depths by Br_2/CH_3OH and 1 HF:3 HNO_3:1 CH_3COOH have been studied. The effect of heat treatment carried out under conditions similar to the epitaxial process of substrates with completely removed damaged layer has been investigated.

Методом рентгеновской топографии определена глубина нарушенного слоя после резки слитков GaSb. Исследовано изменение дефектной структуры подложек после последовательного травления до разных глубин в Br_2/CH_3OH и в 1 HF: 3 HNO_3: 1 CH_3COOH. Прослежено влияние термического отгрева подложек с вполне устраненным нарушенным слоем, который проводился в условиях одинаковых с эпитаксиальным процессом.

1. Introduction

The preparation of the substrate surface and the knowledge of the defect substrate structure are important conditions for obtaining high quality layers by liquid phase epitaxy (LPE) and molecular beam epitaxy (MBE). The semiconductor GaSb is an object of great interest for optoelectronic devices used in fiberoptical connections.

The $A^{III}B^{V}$ compounds have been studied for nearly 25 years but still there are no systematic investigations concerning the influence of technological conditions and surface preparation on the GaSb substrate defect structure.

In the present work the depth of the distorted layer due to diamond disk cutting of wafers is investigated by means of X-ray topography. The change of the defect structure is studied at various depths after successive chemical etching of undoped and Te-doped GaSb substrates. The influence of thermal annealing (in conditions imitating the LPE process) on substrates with fully removed distorted layer is investigated too.

2. Experimental

The investigations were performed on GaSb (111) A and (111) B oriented substrates with area of 4.5 cm^2. The initial dislocation density was $(4 \times 10^3 - 1 \times 10^4)$ cm^2. The slice thickness after diamond disk cutting of Czochralski-grown ingots varies in the range of (500–550) μm. The current carrier concentration in undoped and Te-doped samples was 1.5×10^{17} cm^{-3} and 1.2×10^{18} cm^{-3}, respectively.

After cutting the slices were skimmed and after that washed in distilled water. Next, the samples were etched at room temperature in 3% Br_2/CH_3OH (with 1 min 40 sec duration) or in fast etchant 1 HF:3 HNO_3:1 CH_3COOH (with 5–7 seconds duration). The removed material amount was determined by sample weighting before and after etching. The slice thickness was controlled also by a micrometer. The thermal annealing of the undoped GaSb substrates after the final chemical etching was carried out at $T = 500$ °C for 4 hours, i.e. in conditions used in the LPE process.

X-ray topography was employed for defect observation after every surface removal of the material, as well as after thermal treatment. The topographs were obtained by the anomalous transmission method (BORRMANN) with β-filtered CuK_α radiation. Depending on the sample thickness t, the μt value varies from 24 to 35 (μ is the linear absorption coefficient of X-rays). The (111) A as well as (111) B substrate sides of GaSb wafers were investigated, the corresponding side being the exit of X-rays.

3. Results and discussion

The chemical etching in 3% Br/CH_3OH polishes shiningly the (111) B side; at the same time the (111)A is selectively etched. In order to eliminate the possible influence of this difference on the dislocation pattern, the next removal of material was carried out in 1 HF:3 HNO_3:1 CH_3COOH etchant. The chemical reaction is very fast and difficult to control. But the treatment with this etchant gives shiningly polished surfaces on the both substrate sides.

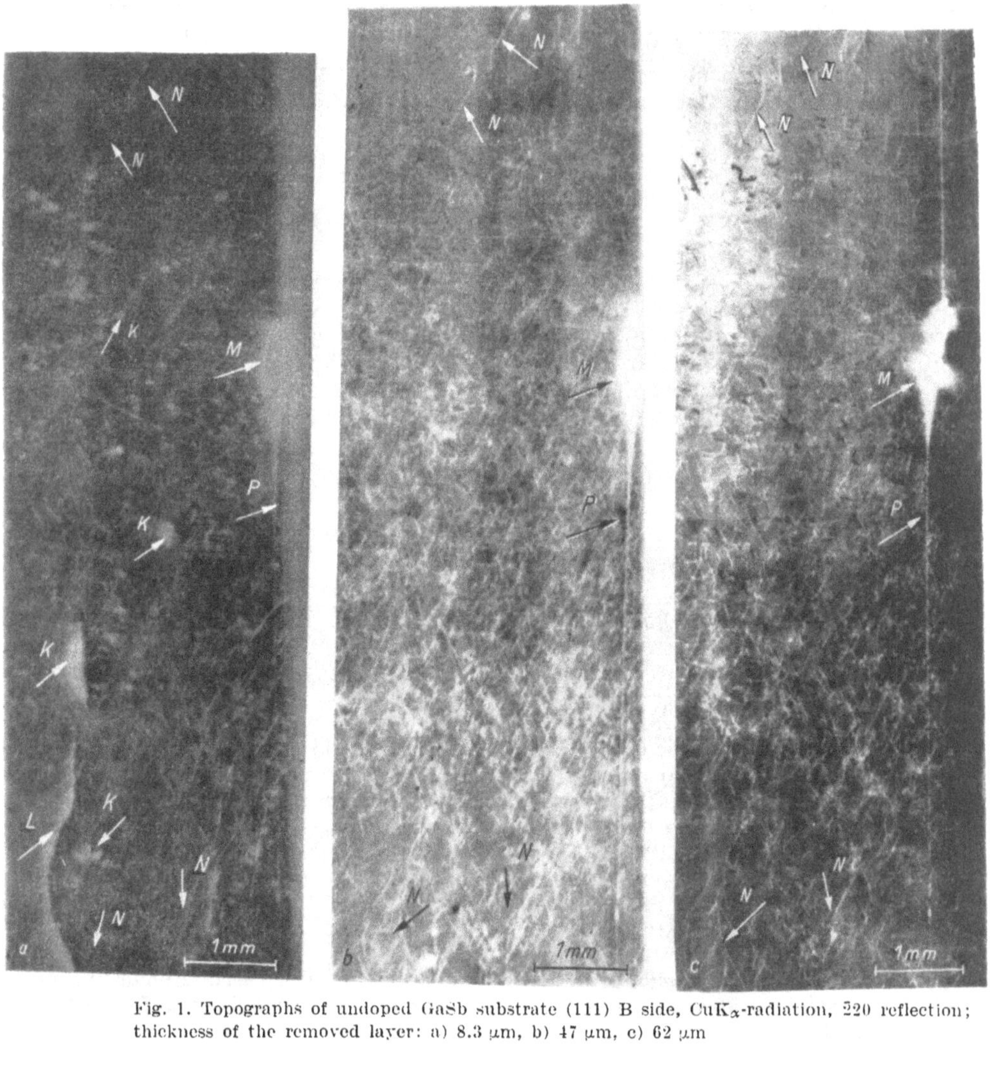

Fig. 1. Topographs of undoped GaSb substrate (111) B side, CuK_α-radiation, 220 reflection; thickness of the removed layer: a) 8.3 μm, b) 47 μm, c) 62 μm

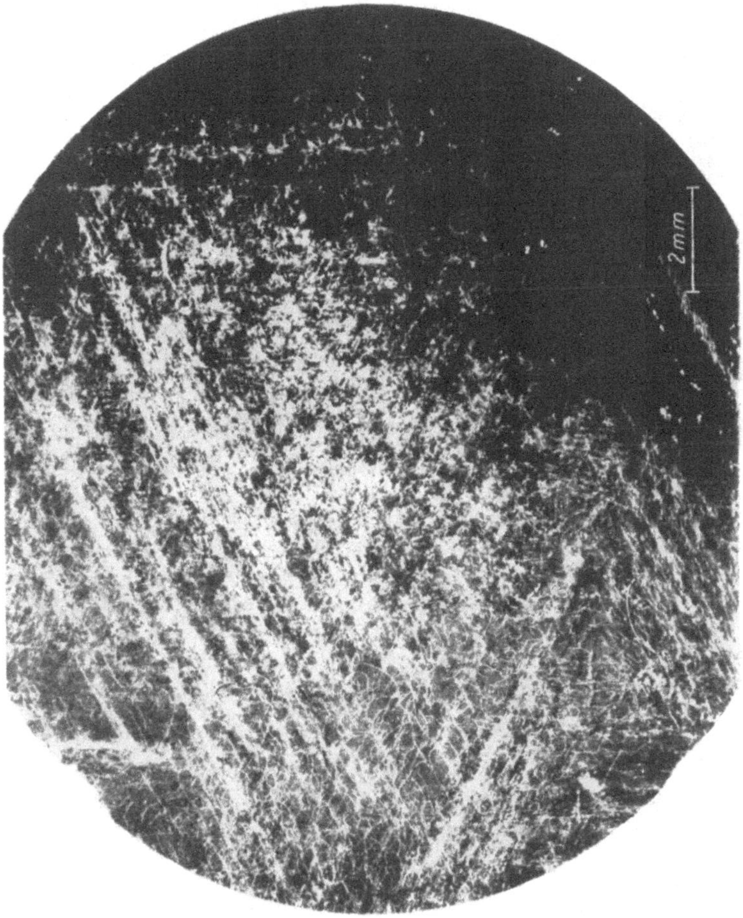

Fig. 3. Topograph of the whole area of undoped GaSb wafer, (111) B side, MoK_α-radiation, $\bar{2}20$ reflection

Fig. 2. Topograph of the same substrate as shown in Figure 1 c), but (111) A side CuK_α-radiation, $\bar{2}20$ reflection

Fig. 4. Topograph of Te-doped GaSb substrate, (111) B side, 220 reflection, thickness of the removed layer 50 μm

The topographs of undoped GaSb received with (111)B side as an exit of X-rays are shown in Figure 1 a—c: (a) after cutting and chemical etching in 3% Br/CH_3OH (the removed layer is 8.3 μm thick); (b) after chemical etching the same substrate in 1 HF:3 HNO_3:1 CH_3COOH (the next removed layer is 47 μm thick); (c) after the same chemical etching (the successive removed layer is 62 μm thick).

On the topographs the defects are imaged with lower diffraction intensity (white) in comparison with the perfect crystal regions. The observed defects may be grouped as follows:

1) white spots labeled by K and L; 2) short fine white lines with complex configuration (labeled by N) which appear as dislocation loops or partial dislocation loops. (MROCZKOWSKI et al.)

Comparing Figures 1a and b (in both figures the same crystal area is presented) one can see that the K- and L-type defects seen in Figure 1a vanish completely after the second etching (Fig. 1b). Therefore the chemical etching at which nearly 47 μm of the surface material are removed leads to the K- and L-type defect elimination, i.e. the creation of these defects must be related to the cutting.

The N-type defects are observed after every etching stage. The dislocation pattern does not change after removal of the next 62 μm of the same substrate. These dislocations are of mixed type according to the contrast image criterion (KUMAR, TAKAGI; KUMAR, MAKACI). It is to be underlined that in the regions of yet removed K-type defects (i.e. after 47 μm removal) the N-type defects are seen distinctly (Fig. 1c). Consequently the N-type defects mut be considered as growth dislocations which are characteristic for the GaSb material.

In Figure 2 the topographs of the same finally etched GaSb substrates is shown, but the (111)A side was an exit for the X-rays. Comparing with Figure 1c it can be seen that there are no significant differences in the dislocation patterns.

A spot of strong local distortion (labeled M) and a prolonged line (labeled P) along [112] direction can be seen in the right hand side of Figures 1 and 2. These imperfections which are observed after every etching step indicate that in all cases the same crystal region is imaged.

The distribution of growth dislocations N is an important question. The topograph received by scanning of the whole wafer area (4.5 cm^2) is shown on Figure 3. As it can be seen a great inhomogeneity of dislocation distribution is observed in Czochralski GaSb ingots.

It is known that in the epitaxial process the substrate undergoes a thermal annealing for a long time (for example 4—6 hours); in many cases the temperature is (500 to 600) °C, which is too high for GaSb with its melting point of 712 °C. To check up the influence of the annealing at the above mentioned conditions, X-ray topographs before and after annealing of the same substrate were taken, changes of the dislocation patterns were not observed; this concerns to both the (111)A and (111)B substrate sides.

Te-doped GaSb substrates were investigated (Fig. 4) at the same conditions as in the case of undoped material. The dislocation pattern on the topograph is different from that of undoped GaSb (Fig. 1c). In the doped material a distinct system of extended 60° straight dislocation lines is observed in addition to the complex configuration dislocation network.

4. Conclusions

As a result of the X-ray topographic investigations the next conclusions can be drawn:

— for preparing substrates free of distortion due to diamond disk cutting, it is necessary to remove about 50 μm from the surface by chemical etching
— there are no significant differences in the (111)A and (111)B side dislocation patterns
— the distribution of the growth dislocations on the slice area is very inhomogenious
— the thermal annealing of undoped GaSb carried out at the conditions usually used in the epitaxial process does not change the defect substrate structure
— in Te-doped GaSb 60° dislocations are observed in addition to the complex configuration dislocation network which is characteristic of the undoped GaSb.

References

Kumar, K., Makaci, M.: Jap. J. Appl. Physics **16** (1977) 957
Kumar, K., Tagaki, M.: Jap. J. Appl. Physics **17** (1978) 299
Mroczkowski, R. S., Witt, A. F., Gotos, H. C.: J. Electrochem. Soc. **115** (1963) 545

(Received June 13, 1987)

Authors' address:
Dr. I. S. Vassilev, Dr. L. D. Pramatarova, Dr. M. B. Baeva
Institute of Solid State Physics, Bulgarian Academy of Sciences
Boulvard Lenin 72
1784 Sofia
Bulgaria

Book Review

B. HALL, A. B. ELLIS (eds)

Chemistry and Structure at Interfaces — New Laser and Optical Techniques

VCH Verlagsgesellschaft, Weinheim, Deerfield Beach, 1986, 351 Seiten, 109 Abbildungen, 16 Tabellen, DM 175.00, ISBN 3-527-26515-5

Die 6 Artikel dieses Buches basieren auf eingeladenen Vorträgen, die auf der Tagung der American Chemical Society im März 1983 in Seattle über Photoprozesse an Festkörperoberflächen gehalten wurden. Die meisten der vorliegenden Review-Artikel, die sich mit den folgenden Themen befassen, enthalten die Literatur bis 1985:

- Untersuchung der Gas-Festkörper- und Metall-Träger-Grenzflächen in Katalysatoren mit Hilfe der kantennahen Röntgenabsorptionsfeinstruktur (NEXAFS oder XANES), vorwiegend unter Nutzung der Synchrotronstrahlung
- Endzustand-aufgelöste Untersuchungen der Wechselwirkung von Molekülen mit Oberflächen mittels Laser-Anregung
- Desorptionsstudien der Kinetik von Oberflächenreaktionen mit Hilfe gepulster Laser-Anregung
- Anwendung der Erzeugung von zweiten Harmonischen für Oberflächenuntersuchungen
- Oberflächen-verstärkte Raman-Spektroskopie (SERS) mit besonderer Betonung von Flüssigkeit-Festkörper-Grenzflächen
- Lumineszenzeigenschaften von Halbleiterelektroden

Die einzelnen Kapitel geben jeweils einen Überblick über die eingesetzten experimentellen Techniken und über die theoretische Grundlage der Methode bei allerdings großen Unterschieden in Umfang und Tiefe der Darstellung. Besonderer Wert wird auf die Anwendung der beschriebenen Methoden gelegt. Insgesamt gesehen, gibt das vorliegende Buch weniger einen vollständigen Überblick über die Arbeit auf diesen Gebieten, als vielmehr an Beispielen orientierte Darstellungen der Fortschritte auf dem sich rasch expandierenden Gebiet der Anwendung von neuen Laser- und optischen Methoden zum Studium von Grenzflächenprozessen.

A. MEISEL

Cryst. Res. Technol. | 23 | 1988 | 2 | 171–177

K. A. Hussain, K. G. Subhadra, K. Kishan Rao

Physics Department, Kakatiya University Warangal India

Habit, Dislocation Densities, and Microhardness of $NaClO_3$-$NaBrO_3$ Mixed Crystals

Results of a systematic study of growth habit, dislocation densities, and microhardness of the $NaClO_3$–$NaBrO_3$ mixed crystals are presented. A drastic change in habit of $NaBrO_3$ for small additions of $NaClO_3$ was observed. The composition dependence of dislocation density for this system is found to be highly non-linear. A non-linear composition dependence of microhardness, with positive deviations from linearity was also observed. This increase in hardness is correlated with the enhanced dislocation density in the mixed crystal.

1. Introduction

Sodium chlorate ($NaClO_3$) and sodium bromate ($NaBrO_3$) are isomorphous and belong to the space group $P\,2_13$ with four molecules per unit cell. These crystals are optically active (Chandrasekaran, Mohanlal) and piezoelectric (Mason); they grow easily from solution. Though isomorphous, $NaClO_3$ and $NaBrO_3$ crystallize with different habits. There is considerable work on the growth characteristics and physical properties of these crystals, particularly of $NaClO_3$, and only few reports on the study of dislocations in these crystals by chemical etching (Kishan Rao, Sirdeshmukh 1978, 1983) and on $NaClO_3$ by X-ray methods (Stepien-Damm et al.; Kito, Kato; Matsunaka et al.; Hooper et al.). $NaClO_3$ and $NaBrO_3$ are known to be completely miscible and form solid solutions over the whole composition range (Chandrasekaran, Mohanlal). A few properties of the mixed crystals of $NaClO_3$ and $NaBrO_3$ have been studied; viz; optical rotation (Chandrasekaran, Mohanlal; Sivaramakrishnan, Arun Kumar 1976); lattice constants (Kleber, Koltermann; Mohanlal) and quadrupole resonance studies (Imaeda). The composition dependence of melting points, lattice constants and coefficient of thermal expansion have been reported by the authors (Subhadra, Hussain 1987 a, b) in earlier communications. This study forms part of a comprehensive programme on the determination of the physical properties of the $Na(ClO_3)_x\,(BrO_3)_{1-x}$ mixed crystals. In the present paper the results of a systematic study of the habit modification, dislocation densities, and microhardness of these mixed crystals are reported.

2. Experimental

Crystals of $NaClO_3$ and $NaBrO_3$ and their mixed crystals over the whole composition range were grown from aqueous solutions, the details of which were reported in an earlier paper (Subhadra, Hussain 1987a). Crystals grown by this method varied in size from a few millimeters to about a centimeter. The composition of crystals was determined polarographically. The studies of dislocations in the mixed crystals were made by chemical etching of the crystals. The etch pits were observed using a Meopta microscope.

For measurement of microhardness a Leitz-Wetzlar microhardness tester fitted with a Vickers pyramid indentor was used. The hardness was determined at various loads in the range from 10 to 75 grams. For each load several indentations were made and the mean value of hardness was taken. This mean value is accurate to within ± 2 kg/mm^2.

3. Results and discussion

3.1. Habit modification

The photomicrographs of some crystals of the $Na(ClO_3)_x(BrO_3)_{1-x}$ system are shown in Figure 1. The $NaBrO_3$ crystal is pyramidal bounded by the (111) planes. In most crystals, there is a sharp intersection of these planes and sharp apices. Sometimes two (111) faces meet in the (110) plane in the form of a narrow edge. The (110) planes meet in a small section of a (100) face. The net result is that the pyramid appears truncated.

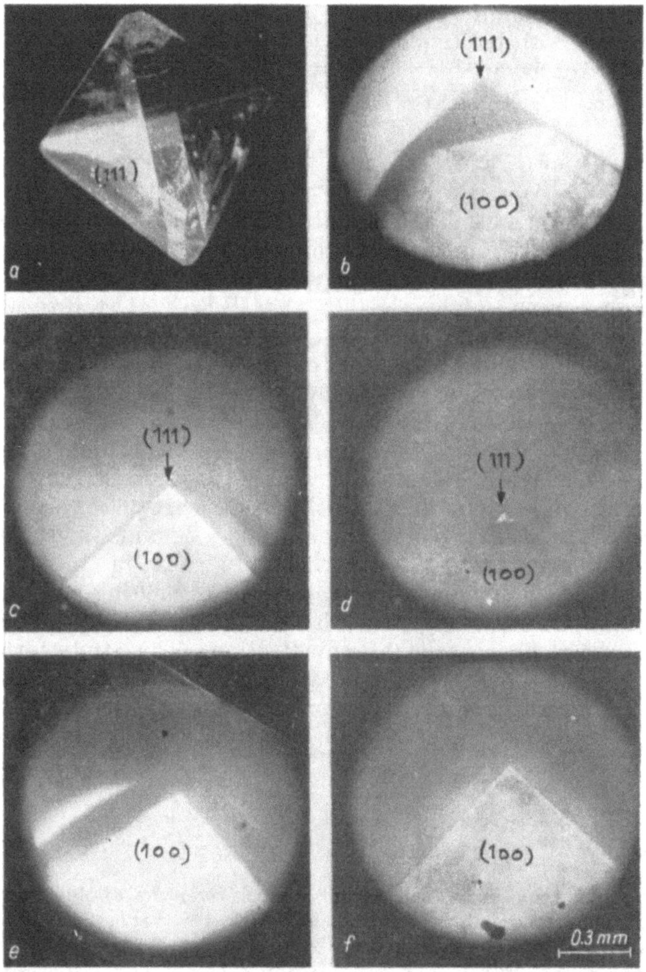

Fig. 1. Photomicrographs showing the habit of crystals of the $Na(ClO_3)_x(BrO_3)_{1-x}$ system. a) Sodium bromate; b) $Na(ClO_3)_{0.015}(BrO_3)_{0.985}$; c) $Na(ClO_3)_{0.05}(BrO_3)_{0.95}$; d) $Na(ClO_3)_{0.105}(BrO_3)_{0.895}$; e) $Na(ClO_3)_{0.13}(BrO_3)_{0.87}$; f) Sodium chlorate

On the other hand, the $NaClO_3$ crystals grow as cubes or plates bounded by the (100) planes. Thus, the habits of $NaBrO_3$ and $NaClO_3$ are entirely different, although the two are isomorphous. CHANDRASEKARAN and MOHANLAL who grew mixed crystals of $NaClO_3$–$NaBrO_3$ comment that the mixed crystals are generally rectangular tablets, but with higher $NaBrO_3$ content (about 70%) some crystals with triangular shape were obtained. But they did not study the habit of the mixed crystals systematically. The change of habit in the $NaClO_3$–$NaBrO_3$ system is shown in photomicrographs (Fig. 1b—f). The linear dimensions of these crystals are nearly the same. As mentioned earlier, in the pure $NaBrO_3$ crystals, the (100) faces are almost non-existent and only the (111) faces are predominantly seen. With the addition of just 1.5 mole percent $NaClO_3$, the habit of $NaBrO_3$ changes drastically, with the (100) faces becoming predominant (Fig. 1a). The (111) faces are still present but with a reduced size. With an increase of $NaClO_3$ content to 10% there is a rapid decrease in the size of the triangular face. For higher $NaClO_3$ content the (111) face is completely absent. Thus, the addition of just about 1 mole per cent $NaClO_3$ has a severe effect on the pyramidal habit of $NaBrO_3$ and with an addition of 10 mole per cent of $NaClO_3$, the (111) faces are completely suppressed. There is as yet no explanation for the difference in the habits of $NaClO_3$ and $NaBrO_3$, nor for the above observation regarding the change in habit in the mixed crystals.

3.2. Etching studies — dislocation densities

Etching studies on sodium chlorate and sodium bromate have been made by KISHAN RAO and SIRDESHMUKH (1978, 1983). In both cases a mixture of acetic acid — formic acid was found to have selective etching action. For $NaClO_3$ a mixture of acetic acid and formic acid (8:1 by volume) gave well defined etch pits on the (100) surface. Acetic acid and formic acid mixture (8:2 by volume) with the addition of 3 mg/cm^3 of cupric nitrate gave well defined etch pits on the (111) faces of sodium bromate. In the present study, the etching action on the $NaClO_3$–$NaBrO_3$ mixed crystals has been studied and the dislocation densities determined.

Freshly grown $NaClO_3$, $NaBrO_3$, and their mixed crystals were used for etching studies. Since very small addition of $NaClO_3$ to $NaBrO_3$ results in the predominance of (100) faces over (111) faces, it was felt that the dislocation densities on the (100) faces of $NaBrO_3$ should be compared with those on the (100) faces of $NaClO_3$ and the mixed crystals. Flat platelets of $NaBrO_3$ with (100) faces were grown by heating the saturated solution to about 60 °C and rapidly cooling it to room temperature.

For both $NaClO_3$ and $NaBrO_3$, a mixture of acetic acid and formic acid (8:1 by volume) gave well defined etch pits on the (100) surface. Figure 2a and 2d show etch pit patterns on the (100) faces of $NaClO_3$ and $NaBrO_3$ crystals. The etch pits on these faces for both the crystals are squares slightly elongated along the diagonal. Because of the point group symmetry of sodium chlorate and sodium bromate, the (100) surface has a diad symmetry. The etch figures on the (100) surfaces of these crystals also exhibit this symmetry. The etch pits are well defined and point bottomed. Occassionally flat bottomed pits are also observed. Most of the pits are of the same size.

Various etchants were tried for the mixed crystals. A mixture of acetic acid and formic acid (8:1 by volume) gave well defined etch pits. Figures 2b and 2c show typical photomicrographs of etched (100) surfaces of some of the mixed crystals. All the crystals, pure as well as mixed, were etched for the same time. In the case of the mixed crystals with molar compositions ranging from 0.3 to 0.75 of $NaBrO_3$, there was difficulty in observing etch pit formation, because the crystal surface became milky when dipped in the etchant. Hence, it was possible to study mixed crystals with molar concentrations of $NaBrO_3$ only upto 0.3 and again for molar concentrations from 0.75

and above. As in the case of the pure end components, the etch pits on the (100) surfaces of all the crystals in the $NaClO_3$–$NaBrO_3$ mixed crystal series are also found to have diad symmetry. The size of the etch pits increased for increased etching time, indicating that the etch pits are formed at dislocation sites.

KISHAN RAO and SIRDESHMUKH (1978, 1983) could not observed any grain boundaries in $NaClO_3$ and $NaBrO_3$. In the present study also grain boundaries have not been observed in the pure crystals (Figs. 2a and 2d). However, it is of interest to note that a number of grain boundaries and, in fact, a network of grain boundaries are observed on the etched surfaces for some of the mixed crystals. SUBBA RAO and HARI BABU also observed that grain boundaries are more in the mixed crystals of KCl–KBr than in the crystals of the pure end components.

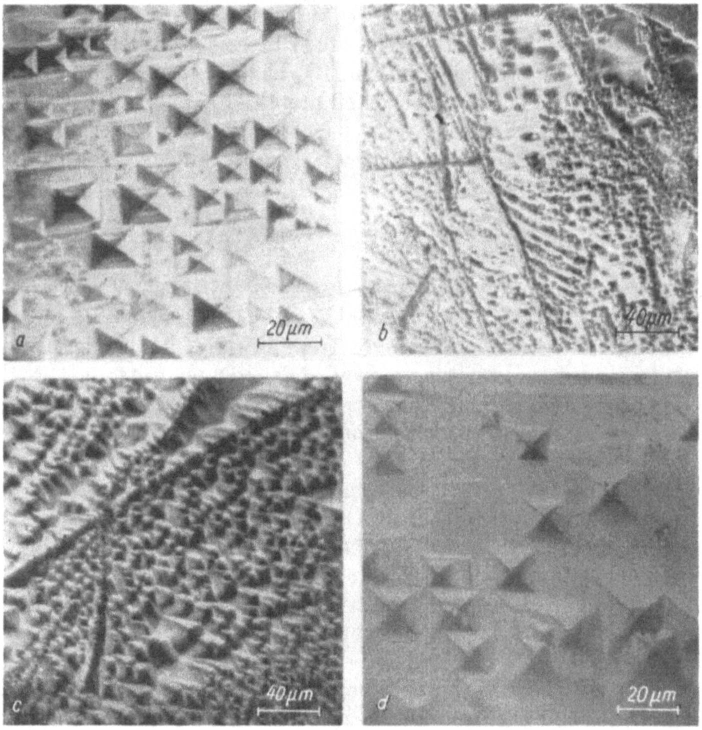

Fig. 2. Etch patterns on the (100) faces of $NaClO_3$, $NaBrO_3$ and their mixed crystals. a) $NaBrO_3$; b) $Na(ClO_3)_{0.07}$ $(BrO_3)_{0.93}$; c) $Na(ClO_3)_{0.7}$ $(BrO_3)_{0.3}$; d) $NaClO_3$

The main objective in the present etching studies is to examine the dislocation densities in mixed crystals vis-a-vis the pure crystals. For this purpose, several photomicrographs were taken for each crystal corresponding to different regions of the surface and also on the equivalent faces of a given crystal, and an average value of the etch pit count was obtained for each crystal. A steep increase in the dislocation density from a value of $16 \times 10^4/cm^2$ for pure $NaClO_3$ to a value of $121 \times 10^4/cm^2$ for a molar fraction of 0.3 of $NaBrO_3$ in the mixed crystal was observed. Similarly, the dislocation density rises from a value of $28 \times 10^4/cm^2$ for $NaBrO_3$ to a value of $100 \times 10^4/cm^2$ for $Na(ClO_3)_{0.25}$ $(BrO_3)_{0.75}$. The composition dependence of dislocation density is shown diagrammatically in Figure 3a. For comparison, the composition dependence of dislocation density in the KCl–KBr system, as observed by SUBBA RAO and HARI

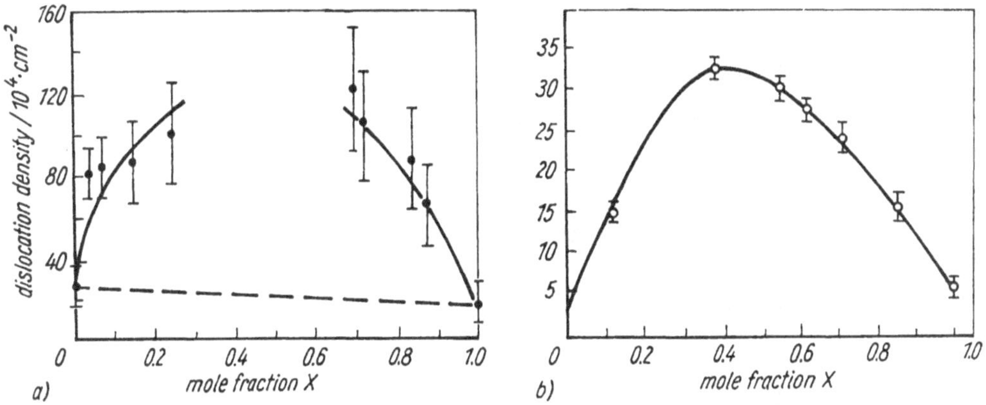

Fig. 3. Plots showing dislocation density as a function of composition (x) for a) $Na(ClO_3)_x(BrO_3)_{1-x}$ system, b) $K(Cl)_{1-x}Br_x$ system (SUBBA RAO, HARI BABU). Dashed line in (a) represents deviations expected from additivity rule

BABU is shown in Figure 3b. It can be seen that the composition dependence of dislocation density in the mixed crystals of $NaClO_3$ and $NaBrO_3$ is similar to that observed in mixed crystals of alkali halides.

3.3. Microhardness

The load variation of hardness for the pure end components and the mixed crystals were determined. It was observed that hardness initially decreases with load and becomes load independent at about a load of 50 grams. The hardness value at the highest load employed (75 grams) is taken as the representative value for the crystal. These hardness values are given in Table (1) along with values for the pure crystals obtained by earlier workers. The present values for the pure crystals are nearer to the values reported by HAUSSÜHL than to the values reported by KISHAN RAO and SIRDESHMUKH (1983).

The composition dependence of hardness is shown in Figure 4. The variation with composition is nonlinear. The values of ΔH_v (the deviation of the experimental hardness values from that expected from a linear composition dependence) are shown as a function of composition in Figure 5.

It can be seen that this deviation increases with composition, reaches a maximum at about the equimolar composition and then, decreases again. Qualitatively, this

Table 1
Values of microhardness (H_v) for $Na(ClO_3)_x(BrO_3)_{1-x}$ crystals

x	H_v (kg/mm²)		
	present work	HAUSSÜHL	KISHAN RAO SIRDESHMUKH (1983)
0.00	215	195	136
0.05	220		
0.10	218		
0.45	182		
0.56	174		
0.80	148		
1.00	117	123	107

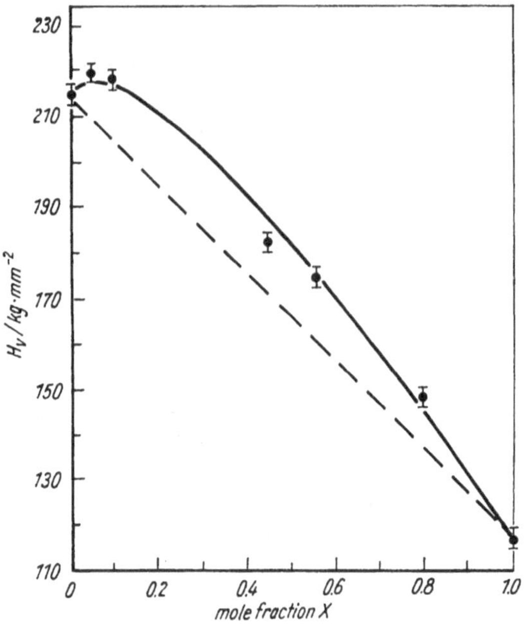

Fig. 4. Plot of hardness (H_v) against composition (x) for the $Na(ClO_3)_x(BrO_3)_{1-x}$ system. Dashed line represents the deviation expected from additivity rule

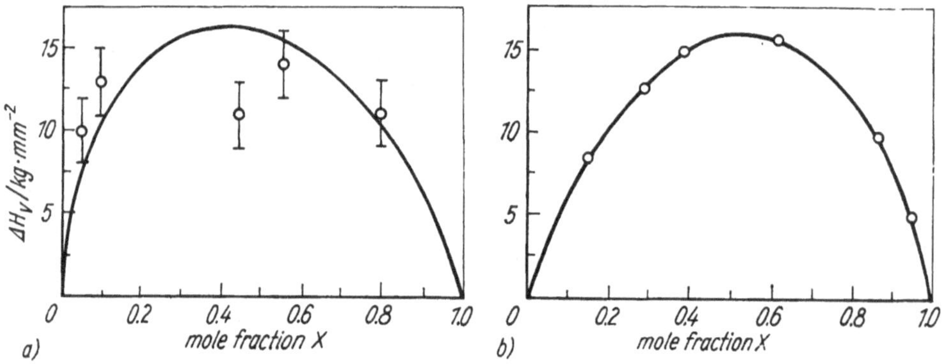

Fig. 5. Plot of ΔH_v against composition (x). a) $Na(ClO_3)_x(BrO_3)_{1-x}$ system; b) KCl_xBr_{1-x} system (SUBBA RAO, HARI BABU)

deviation is very similar to the deviation observed in the case of KCl–KBr by SUBBA RAO and HARI BABU (also shown in Figure 5) although the percentage increase in hardness in the equimolar region compared to the mean hardness of the end members is much more in the KCl–KBr system (nearly 200 percent) than in the $NaClO_3$–$NaBrO_3$ system (nearly 10 percent).

3.4. Correlation between microhardness and dislocations

SUBBA RAO and HARI BABU (1978) have determined the microhardness of KCl–KBr, KCl–KI, and KCl–NaCl mixed crystal systems. In all these systems they observed that the microhardness varied with composition in a highly non-linear fashion, the values for mixed crystals exceeding the values for the end members. They pointed out that the microhardness of mixed crystals has two components, the first is the lattice component which varies according to r^{-m}, where r is the interionic distance and m the Born index. The value of m is about 8 for alkali halides and this gives an almost linear deviation with composition for the lattice hardness. The second component is due to the size effect. The presence of two types of ions of different size at equivalent lattice points creates strains in the lattice which in turn give rise to dislocations. These extra dislocations hinder the mobility of other dislocations and cause an apparent increase in hardness. This effect, understandably, increases with composition from either side for the solid solution. The observed variation of hardness for alkali halide mixed crystals has been explained by SHRIVASTAVA on the basis of the KATAOKA-YAMADA model. The composition dependence of microhardness of CaF_2–SrF_2 mixed crystals was determined by PLENDL et al. The observed results were explained by them in terms of lattice anharmonicity.

As discussed in sections 3.2 and 3.3 an increase is observed in the dislocation density with composition from either end of the series for the $NaClO_3$–$NaBrO_3$ system, and the composition dependence of hardness shows slight deviations from linearity. The composition dependence of ΔH_v (Fig. 5) and of the dislocation density are similar to what has been observed in the case of the alkali halide mixed crystals. Hence, the simple interpretation given by SUBBA RAO and HARI BABU in the case of alkali halide mixed crystals that the enhanced dislocation density hinders the motion of dislocations and causes an enhancement of microhardness, should account for the observed composition variation of dislocation density and microhardness of the $NaClO_3$–$NaBrO_3$ system also.

Thanks are due to Prof. D. B. Sirdeshmukh for suggestions and encouragement.

References

CHANDRASEKARAN, K. S., MOHANLAI, S. K.: Pramana 7 (1976) 152
HAUSSÜHL, S.: Phys. Kondens Mater. 3 (1964) 139
HOOPER, R. M., ROBERTS, K. J., SHERWOOD, J. N.: J. Mat. Sci. 18 (1983) 81
IMAEDA, Y.: J. Phys. Soc. Japan 15 (1960) 1699
KATAOKA, T., YAMADA, T.: Japan J. Appl. Physics 16 (1977) 1129
KISHAN RAO, K., SIRDESHMUKH, D. B.: J. Cryst. Growth 44 (1978) 533
KISHAN RAO, K., SIRDESHMUHK, D. B.: Cryst. Res. Technol. 18 (1983) 1125
KITO, I., KATO, N.: J. Cryst. Growth 24/25 (1974) 544
KLEBER, W., KOLTERMANN, M.: Wissenschaftliche Z. der Humboldt-Universität zu Berlin, (1958) p. 625
MATSUNAKA, M., KITAMURA, M., SUNAGAWA, I.: J. Cryst. Growth 48 (1980) 425
MASON, W. P.: Phys. Rev. 70 (1946) 529
MOHANLAL, S. K.: Ph. D. Thesis, Madurai University, Madurai 1971
PLENDL, J. N., GIELISSE, P. J., MANSUR, L. C., MITRA, S. S., SMAKULA, A., TARTE, P. C.: Applied Optics 10 (1971) 1129
SHRIVASTAVA, U. C.: J. Appl. Phys. 51 (1980) 1510
SIVARAMAKRISHNAN, V., ARUN KUMAR, K. A.: J. Phys. Chem. Solids 37 (1976) 799
STEPIEN-DAMM, J. A., KUCHARCZYK, D., URBANOWICZ, E., LUKASZEWICZ, K.: Bulletin de l' Academie Polonsaise des Sciences XXIII, 1975, p. 985
SUBBA RAO, U. V., HARI BABU, V.: Pramana 11 (1978) 149
SUBHADRA, K. G., HUSSAIN, K. A.: Cryst. Res. Technol. 22 (1987) a 827
SUBHADRA, K. G., HUSSAIN, K. A.: Cryst. Res. Technol. 22 (1987) b 1165

(Received May 27, 1987)

Authors' address:

K. A. HUSSAIN, Dr. K. G. SUBHADRA, K. KISHAN RAO
Department of Physics, Kakatiya University
Warangal — 506009, India

Book Review

M. A. VAN HOVE, W. H. WEINBERG, C.-M. CHAN

Low-Energy Electron Diffraction

Springer-Verlag Berlin Heidelberg, 1986, 603 pages, 220 figs., 20 tables. DM 124.00. ISBN 3-540-16262-3

The monography "Low-Energy Electron Diffraction" (LEED), written by three well-known specialists in this field and published within the Springer series in Surface Sciences, is a unique representation of this powerful method. According to the given subtitle "Experiment, Theory and Surface Structure Determination" the whole discipline of low-energy electron diffraction is covered.

The authors state that similar to solid state physics, where the detection of the correlation between the microstructure and the properties of bulk materials is of the utmost importance, in surface science there is an exigence of elucidating the connection between surface structure and surface properties. With respect to crystalline surfaces LEED is the key method to allow a reliable characterization of the relative position of atoms, thus being the basis for attaining exprimental results in surface crystallography.

Starting with the description of the historical development of LEED the authors describe practical aspects for carrying out LEED experiments (instrumentation, data acquisition) and for gaining structure information by carefully interpreting diffraction patterns. It follows a very clear treatment of the kinematical theory of LEED and of the dynamical LEED theory.

The field of application of the method is dealt with in detail. The authors point to the results of structural analyses with respect to clean unreconstructed and reconstructed surfaces, resp., and to adsorbed atomic and molecular layers, resp. Special emphasis is placed on the detection and explanation of two-dimensional orderdisorder phase transitions, e.g. given by the interaction of hydrogen with well-defined metal surfaces (Ni, Pd, Fe). As further fields of the application of LEED, chemical reactions at surfaces are mentioned as well as the island formation of a species. Finally, the authors point to the future trend with respect to the LEED experimentation and theory and to the progress in structure determination.

Besides the complete description of the possibilities and limitations of the LEED technique the book shows the importance of this method within other surface-sensitive techniques (for comparison, a valuable table is added). With respect to the results of LEED investigations, a detailed reference list is given as well as a table of surface structures.

The clearly written and well-illustrated book should be of great interest for researchers working in surface science and surface techniques, e.g. in the field of catalysis, corrosion, coatings, adhesion, lubrication, and last not least in microelectronics. On the one hand, for a scientist working in the general field of surface problems the book is an excellent introduction showing the power of the LEED technique; on the other hand, for the specialist being already familiar with low-energy electron diffraction, in a concise manner the book offers and updated representation of the field and of experimental and theoretical details, supplemented by a valuable list of about 700 references.

There is no doubt that VAN HOVE, WEINBERG and CHAN's book will find many readers all over the world and that it will be an essential stimulation in surface science in general, and in the application of the LEED technique in particular.

J. HEYDENREICH

Cryst. Res. Technol. | 23 | 1988 | 2 | 179–184

B. MAJUMDAR, B. N. DUBEY, B. P. CHANDRA

Department of Physics, Government Science College, Raipur, India and
Department of Mathematics, Government Science College, Raipur, India, and
Department of Physics, Rani Durgawati University, Jabalpur, India

An Estimation of Fracture Stress in a Brittle Solid

The present paper reports for the first time an exact analytic expression that can be used for obtaining fracture stress σ_F, when fracture occurs in the elastic limit and also when fracture occurs after general yield ($\sigma_F > \sigma_Y$) in a brittle solid. The present paper may provide a new basis to fracture-safe design of brittle materials.

1. Introduction

When a solid is placed under tensile stress by stretching forces, its molecular bonds resist the elongation. The ratio of the elongation to the initial length is called strain. If the strain is less than 1%, the bonds can be regarded as tiny springs linking the molecules. If the stress is removed, the solid regains its initial length as the springs contact. Under a larger strain a ductile solid first is distorted and then breaks.

The mechanism of breaking is by means of cracks that form and then extend through the solid. Fracture is an inhomogeneous form of deformation which can be viewed on different levels (TETELMAN). On an atomistic level it occurs by the breaking of atomic bonds, either perpendicular to a plane (cleavage) or across a plane (shear). On a microscopic level cleavage occurs by the formation and propagation of microcracks and the separation of grains along cleavage planes. Shear fracture (rupture) usually occurs by the formation of voids within grains and the separation of material between the voids by intense shear. On a macroscopic level cleavage occurs when a cleavage crack spreads essentially perpendicular to the axis of maximum tensile stress. Shear fracture occurs when a fibrous crack advances essentially perpendicular to the axis of maximum tensile stress (normal rupture) or along a plane of maximum shear stress (shear rupture).

In a perfectly elastic solid, fracture occurs when the stress level at the tip of a preinduced flaw reaches the theoretical cohesive stress and a sufficient amount of work is done to break atomic bonds and create new surfaces. Since the stress concentration factor of an elastic crack, is proportional to $c^{1/2}$, the nominal fracture stress is proportional to $(\gamma_s/c)^{1/2}$ as predicted by GRIFFITH equation. This estimation only applies for completely brittle and completely elastic solids where (i) the elastic limit or yield strength σ_Y is much greater than the cohesive stress (approximately $E/10$) and (ii) there is also no microplastic strain occurring near the crack tip.

Once the crack has started to spread, its length $2c$ increases. If the tip radius ϱ remains constant or decreases as the crack extends or if c/ϱ remains constant, then the crack will spread across the material and cause complete fracture of the part, provided that σ is maintained constant during crack propagation. Therefore, σ_F is the applied stress required to cause complete facture (i.e. the fracture stress).

When the yield stress of a material is less than $E/10$ (i.e. in a ductile material) or in the vicinity of a stopped crack in a partially brittle one, plastic flow occurs near the crack tip and the stress level in the plastic zone is less than $E/10$. Consequently, the crack cannot advance directly as an elastic "Griffith" crack and fracture is then plastically induced.

Resistance to brittle failure (the degree of difficulty with which cracks propagate through a stressed material) is measured quantitatively as fracture toughness (KENT BOWEN) and has vital role in fracture-safe design (PELLINI, PUZAK). This requires a complete knowledge of fracture stress σ_F in low-yield strength material (in which $\sigma_F > \sigma_Y$) and in high-yield strength material (in which $\sigma < \sigma_F$). The present paper reports for the first time an exact analytic expression that can be used for obtaining fracture in a brittle solid.

2. Theoretical

Suppose, we have a crystal of cubic shape having volume V, and it is being compressed uniaxially at a particular strain compression rate. After a certain compressive strain, the new cracks will be formed in the crystal. If the crystal is fractured into N crystallites, then the area of newly created surfaces will increase. In real fracture, all the crystallites may not be of the same size. Thus, if we go to the reality it will be very difficult to have an idea about the area of the newly created surfaces. For the convenience of the estimation, let us assume that the average volume of a crystallite is (V/N) and thereby the average surface area of a crystallite is $6(V/N)^{2/3}$. Now, the area S of the newly created surfaces after the creation of N crystallites may be expressed as

$$S = 6N\left(\frac{V}{N}\right)^{2/3} - 6V^{2/3},$$

or

$$S = 6V^{2/3}(N^{1/3} - 1)\,. \tag{1}$$

It should be noted that when the crystal is crushed into crumbled masses, then many thousands of crystallites are produced. In such cases, the above approach may be correct from the statistical point of view. Since stress is more fundamental quantity rather than strain, in other words, stress controls the strain in solids. It will be more appropriate to change the relations related to strain (CHANDRA, DUBEY) in terms of the stress.

From equation (1), the rate of creation of new surfaces dS with the stress $d\sigma$ of the crystal may be given by

$$\frac{dS}{d\sigma} = \frac{dS}{dN} \cdot \frac{dN}{d\sigma},$$

or

$$\frac{dS}{d\sigma} = 2V^{2/3}N^{-2/3}\frac{dN}{d\sigma}. \tag{2}$$

Let dN be the number of crystallites formed during the change of strain from ε to $\varepsilon + d\varepsilon$, then dN may be written as

$$dN = M\,d\varepsilon\,, \tag{3}$$

where M is the multiplication factor which when multiplied with the strain, the product gives the increase in the number of crystallites.

The multiplication factor M increases with increasing number of crystallites, and it should also change with the deformation of the crystal. Thus, taking these two factors into account, the change dM during the change of strain from ε to $\varepsilon + d\varepsilon$ may be given by

$$dM = \alpha\, dN + \beta\, d\varepsilon$$

or

$$dM = \alpha M\, d\varepsilon + \beta\, d\varepsilon\,, \tag{4}$$

where α and β are constants.

From equation (4) $\dfrac{dM}{M + \beta/\alpha} = \alpha\, d\varepsilon$. At this stage, strain $d\varepsilon$ can be replaced in terms of stress, by assuming that volume and stress applied to the sample, may be given by $V = Al$, and $P = A\sigma$, where A is the uniform cross-sectional area and l is the initial length of the sample. Assuming that volume remain unaltered during deformation, we have

$$A dl + l\, dA = 0\,. \tag{5}$$

Moreover, taking the differential of the relation $P = A\sigma$ shows that $P = P_{\max}$ when

$$dP = \sigma\, dA + A\, d\sigma = 0 \tag{6}$$

or

$$\frac{d\sigma}{\sigma} = -\frac{dA}{A} = d\varepsilon\,. \tag{7}$$

Using relation (7), equation (4) leads to

$$\frac{dM}{M + \beta/\alpha} = \alpha \frac{d\sigma}{\sigma}$$

$$\text{or} \log\,(M + \beta/\alpha) = \alpha \log \sigma + \log c \tag{8}$$

where c is constant.

The multiplication factor M is zero till $\sigma = \sigma_F$, the fracture stress. Therefore equation (8) may be written as

$$\log\,(M + \beta/\alpha) = \alpha \log \sigma + \log \beta/\alpha - \log \sigma_F^\alpha$$

or

$$\log\,(M + \beta/\alpha) - \log \beta/\alpha = \alpha \log \sigma - \log \sigma_F^\alpha$$

or

$$\frac{M + \beta/\alpha}{\beta/\alpha} = \left(\frac{\sigma}{\sigma_F}\right)^\alpha$$

or

$$M = \left(\frac{\beta}{\alpha}\right)\left[\left(\frac{\sigma}{\sigma_F}\right)^\alpha - 1\right]$$

or

$$M = M_0\left[\left(\frac{\sigma}{\sigma_F}\right)^\alpha - 1\right] \tag{9}$$

where

$$M_0 = \beta/\alpha\,.$$

From equation (3) and (7), dN may be written as

$$dN = M\, d\varepsilon = M_0\left[\left(\frac{\sigma}{\sigma_F}\right)^\alpha - 1\right] d\varepsilon$$

or

$$dN = M_0\left[\left(\frac{\sigma}{\sigma_F}\right)^\alpha - 1\right] \frac{d\sigma}{\sigma} \tag{10}$$

or

$$N = \frac{M_0}{\alpha}\left(\frac{\sigma}{\sigma_F}\right)^\alpha - M_0 \log \sigma + c_1. \tag{11}$$

When the applied stress reaches to fracture stress, σ_F, i.e. when fracture just starts, the number of crystallites N is equal to one for $\sigma = \sigma_F$. Thus

$$c_1 = \frac{M_0}{\alpha}\left[\frac{\alpha}{M_0} + \alpha \log \sigma_F - 1\right]. \tag{12}$$

After substituting this value of c_1 in equation (11), we have

$$N = \frac{M_0}{\alpha}\left[\left(\frac{\sigma}{\sigma_F}\right)^\alpha - \alpha \log\left(\frac{\sigma}{\sigma_F}\right) - 1\right] + 1. \tag{13}$$

Substituting the values of N and $\frac{dN}{d\sigma}$ from equation (13) and (11) in equation (2) leads to the result

$$\frac{dS}{d\sigma} = \frac{2M_0^{1/3}V^{2/3}\alpha^{2/3}}{\sigma}\left[\left(\frac{\sigma}{\sigma_F}\right)^\alpha - 1\right]\left[\left\{\left(\frac{\sigma}{\sigma_F}\right)^\alpha - 1\right\} - \log\left(\frac{\sigma}{\sigma_F}\right)^\alpha + \frac{\alpha}{M_0}\right]^{2/3}. \tag{14}$$

In order to estimate the value of the fracture stress σ_F, let us find the maximum or minimum values of $\frac{dS}{d\sigma}$. The maximum or minimum values of $\frac{dS}{d\sigma}$ are given by

$$f'(\sigma) = \frac{d^2S}{d\sigma^2} = 0. \tag{15}$$

This condition when subjected to equation (14) gives the following equation

$$\left[\frac{\alpha\sigma^\alpha}{(\sigma^\alpha - \sigma_F^\alpha)^2} - \frac{1}{(\sigma^\alpha - \sigma_F^\alpha)}\right]\left[\left\{\left(\frac{\sigma}{\sigma_F}\right)^\alpha - 1\right\} - \log\left(\frac{\sigma}{\sigma_F}\right)^\alpha + \frac{\alpha}{M_0}\right] = \frac{2}{3}. \tag{16}$$

Since the complete solution of equation (16) is difficult, we can consider two cases: one for $\sigma \ll \sigma_F$ and the other for $\sigma \gg \sigma_F$.

Case 1: When $\sigma \ll \sigma_F$ i.e. when fracture occurs in the elastic limit.

Equation (16) may also be written as

$$\left[\frac{\alpha\sigma^\alpha}{(\sigma_F^\alpha)^2\left[1 - \left(\frac{\sigma}{\sigma_F}\right)^\alpha\right]^2} + \frac{1}{(\sigma_F^\alpha)\left[1 - \left(\frac{\sigma}{\sigma_F}\right)^\alpha\right]}\right] \times$$
$$\times \left[\left\{\left(\frac{\sigma}{\sigma_F}\right)^\alpha - 1\right\} - \log\left(\frac{\sigma}{\sigma_F}\right)^\alpha + \frac{\alpha}{M_0}\right] = \frac{2}{3},$$

or

$$\frac{1}{(\sigma_F)^\alpha}\left[\alpha\left(\frac{\sigma}{\sigma_F}\right)^\alpha\left\{1 - \left(\frac{\sigma}{\sigma_F}\right)^\alpha\right\}^{-2} + \left\{1 - \left(\frac{\sigma}{\sigma_F}\right)^\alpha\right\}^{-1}\right]\left[\frac{\alpha}{M_0} - \log\left(\frac{\sigma}{\sigma_F}\right)^\alpha - 1\right] = \frac{2}{3},$$

or

$$\frac{1}{(\sigma_F)^\alpha}\left[\alpha\left(\frac{\sigma}{\sigma_F}\right)^\alpha\left\{1 + 2\left(\frac{\sigma}{\sigma_F}\right)^\alpha + \ldots\right\} + \left\{1 + \left(\frac{\sigma}{\sigma_F}\right)^\alpha + \ldots\right\}\right] \times$$
$$\times \left[\frac{\alpha}{M_0} - \log\left(\frac{\sigma}{\sigma_F}\right)^\alpha - 1\right] = \frac{2}{3},$$

or

$$\frac{1}{(\sigma_F)^\alpha}\left[1 + \left(\frac{\sigma}{\sigma_F}\right)^\alpha(\alpha + 1) + \ldots\right]\left[\frac{\alpha}{M_0} - \log\left(\frac{\sigma}{\sigma_F}\right)^\alpha - 1\right] = \frac{2}{3}.$$

When $\sigma \ll \sigma_F$ or $\left(\frac{\sigma}{\sigma_F}\right) \ll 1$ we may neglect the quantity $\frac{\sigma}{\sigma_F}(\alpha+1)$ as compared to 1. This leads to

$$\frac{1}{(\sigma_F)^\alpha}\left[1 + \alpha \log\left(\frac{\sigma}{\sigma_F}\right) - \frac{\alpha}{M_0}\right] = -\frac{2}{3},$$

which gives the following expression for fracture stress σ_F when $\sigma \ll \sigma_F$

$$\sigma_F = \left\{-\frac{3}{2}\log\left[\frac{1}{(\sigma_F)^\alpha}\exp\left[1 + \alpha \log \sigma - \frac{\alpha}{M_0}\right]\right]\right\}^{1/\alpha}. \quad (17)$$

Case II: When $\sigma \gg \sigma_F$ i.e. when fracture occurs after general yield.

Subjected to this condition, equation (16) yields the result for fracture stress σ_F as follows

$$\sigma_F = \left[\exp\left\{\frac{2}{3}\frac{\sigma^\alpha}{(\alpha - 1)} - \left(\frac{\sigma}{\sigma_F}\right)^\alpha - \frac{\alpha}{M_0}\right\}\right]^{1/\alpha}. \quad (18)$$

3. Importance

We know that the value of the toughness is rarely constant for a given material. It is strongly dependent on the mode of fracture initiation and this, in turn, depends on size of the structure (particularly thickness), environmental conditions, strength level, etc. Consequently, in fracture-safe design, one requires that the toughness be estimated on the basis of the worst possible conditions that a material will be exposed to during service.

The theoretical calculation presented here for the estimation of fracture stress in brittle materials would possibly become difficult to determine especially in structures having complicated shapes and subject to complex loading. Despite these difficulties, we have made an attempt for the first time as far as the literature survey shows for the estimation of fracture stress σ_F, the knowledge of which will be extremely useful for producing fracture-safe designs in high strength materials at all temperatures and in low and medium strength materials at low temperatures. However, it must be emphasized, that the above approach for producing fracture-safe designs is only as good as (i) the non-destructive testing procedure (Entwistle; Bradfield; Lamble) that is used to detect the flaws and their slow growth; (ii) the stress analysis procedures (Hetenyi; Frocht), that are used to determine the value of σ at all critical points in the structure; when combined along with the analytical aspects of the fracture mechanics will furnish more complete knowledge in this particular field of fracture-safe design in brittle materials.

The authors are indepted to Professors P. C. Kamra and R. S. Matharu, as well to M. Elyas for their constant support and encouragement.

References

Bradfield, G.: Physical Examination of Metals, London 1960, pp. 559, 605
Chandra, B. P., Dubey, B. N.: Cryst. Res Technol. **20** (1985) 997
Entwistle, K. M.: Physical Examination of Metals, London 1960, pp. 487
Frocht, M. M.: Photoelasticity, New York 1941
Griffith, A. A.: Proc. Int. Congr. Appl. Mech. **55** (1924)
Hetenyi, M.: Handbook of Experimental Stress Analysis, New York 1950

KENT BOWEN, H.: Advanced Ceramics, **255** (1986) 146
LAMBIE, J. H.: Principles and Practice of Non-Destructive Testing, New York 1963
PELLINI, W. S., PUZAK, P. P.: NRL Report 5920 (March 1963)
PELLINI, W. S., PUZAK, P. P.: NRL Report 6030 (November 1963)
PELLINI, W. S., PUZAK, P. P.: Trans. ASME (J. Eng. Power) (1964), p. 429
TETELMAN, A. S., MCEVILY, A. J., Jr.: Fracture of Structural Materials, 1967

(Received May 27, 1987)

Authors' addresses:

Dr. B. MAJUMDAR, Ass. Professor
Dept of Physics, Government Science College
Raipur (M.P.), 492002, India

B. N. DUBEY
Dept of Mathematics, Government Science College
Raipur, 492002, India

Dr. B. P. CHANDRA
Dept of Physics, Rani Durgawati University
Jabalpur, 482001, India

Cryst. Res. Technol.	23	1988	2	185–193

K. B. R. Varma, A. R. Raju, K. J. Rao

Materials Research Laboratory, Indian Institute of Science, Bangalore, INDIA

Physico-chemical Properties of Pure and X-ray Irradiated Single Crystals of KAP[1])

Single crystals of potassium hydrogen phthalate (KAP) have been grown by slow evaporation method from aqueous solutions. Thermal analyses indicate that KAP crystals decompose into phthalic anhydride and KOH around 520 K. Electrical properties of single crystals of KAP have been studied along with the effect of X-ray irradiation of the crystals. The electrical transport appears to be associated with tunneling of protons. The irradiated crystal exhibits lower dielectric constant and higher ac conductivity.

1. Introduction

Single crystal of potassium hydrogen phthalate, commonly known as potassium acid phthalate (KAP) has wide technological applications because of its high piezoelectric coefficient and mica-like layered structure. Its piezoelectric coefficient is similar to that of lead zirconate titanate. KAP single crystals are easily cleaved into very thin layers in the plane perpendicular to the crystallographic *c*-axis along which it has a large '*d*' spacing of 13.26 Å. This property has indeed been exploited by Jones et al. in the production of crystal analyzers for soft X-ray spectrometers.

In view of its technological importance, several studies of KAP have been reported in the literature. In-situ observations of growth aspects of KAP single crystals have been made by van Enckevort et al. and Jetten et al. Distler et al. have examined the mechanism of crystal layer growth using decoration technique and electron microscopic studies. Koptsik et al., Chumakov et al., and Gavrilova have reported that these crystals exhibit both piezo- and pyroelectric effects. According to Belyaev et al. KAP melts at 290 °C and is stable upto 400 °C whereas it is reported to be unstable (Lange's hand book of chemistry, 13th ed., 1986) and is supposed to decompose well before melting. Thus many of the reported studies require a critical examination.

KAP single crystals possess a promising combination of properties like high birefringence and high piezoelectric coefficients which can be exploited in making devices like electro- and acoustooptic modulators. A detailed investigation of the physico-chemical properties of KAP crystals is therefore considered as essential. Also, to the best of the knowledge of the authors, no systematic studies of the electrical properties of KAP single crystals have been reported in the literature.

In this communication, we present several physico-chemical studies of KAP single crystals such as dielectric constants and ac conductivities (along crystallographic *c*-direction as a function of both temperature and frequency), chemical stability, infrared transmission, and X-ray diffraction. Since KAP is an organic salt, we considered it interesting to investigate the effect of X-ray irradiation on these crystals. This

---

[1]) Contribution No. 75 from Materials Research Laboratory.

paper also reports results of our preliminary studies on electrical conductivity, dielectric, and optical properties of X-ray irradiated crystals of KAP.

2. Experimental

KAP single crystals used in this investigation were grown from aqueous solutions by slow evaporation at 300 K. X-ray powder pattern of the material (powder of the grown crystals) was obtained on a Philips PW 1050/70 diffractometer. The chemical stability (decomposition temperature etc.) of KAP was investigated by thermogravimetric and differential thermal analysis (TGA and DTA) in the temperature range of 300 to 675 K, using an Ulvac Thermal Analyzer, model TA 1500, at a heating rate of 10 K/min. Infraied transmission spectra of the product of decomposition (sublimate) of KAP and phthalic anhydride powder (Analar) were recorded on a Perkin Elmer 599 Infrared Spectrophotometer in the 200—4000 cm^{-1} range at room temperature in KBr pellets. Capacitance of KAP single crystal was measured as a function of both temperature (300—425 K) and frequency (1—100 kHz) with a signal strength of 8 V rms using three terminal General Radio GR 1615-A capacitance bridge coupled externally with GR 1316 oscillator and GR 1232 tuned amplifier with a null detector. The capacitance of KAP single crystal at 1 MHz was measured seperately, using a Boonton (USA) capacitance meter model 72 BD. The stray capacitance introduced by leads was minimized by reducing the lengths of the connecting leads and by providing adequate shielding. Dielectric constants were evaluated after correcting for the residual stray capacitance. The ac conductivity measurements were carried out on a General Radio (USA) impedance bridge model 1608 A. A single crystal of KAP was irradiated by filtered X-rays (CuK_α radiation, 30 kV, 20 mA) at room temperature for about 8 hours. Electron spin resonance spectrum of irradiated crystal was obtained on varian E-109 X-band ESR spectrometer.

3. Results and discussion

A typical single crystal of KAP of dimensions roughly 25 mm (length) × 25 mm (breadth) × 12 mm (thickness) is shown in Figure 1. The crystal structure was confirmed by comparing its X-ray powder pattern with standard KAP powder pattern from literature. The crystallographic axes (*a*, *b*, *c*) were labelled based on its morphology and labelling scheme proposed by GROTH. The crystal plates were cleaved perpendicular to the *c*-axis and were ground and polished.

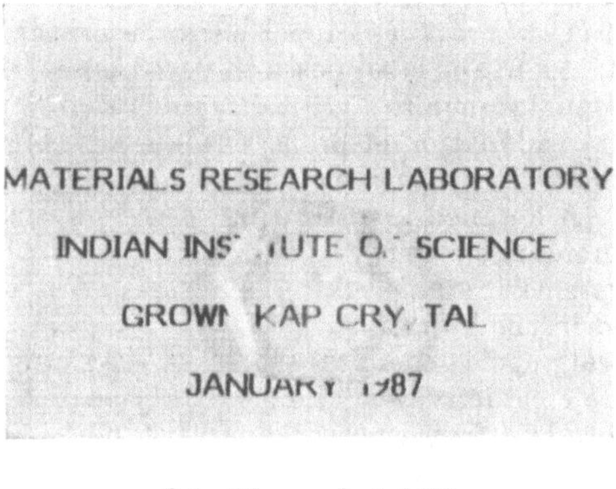

Fig. 1. Potassium hydrogen phthalate single crystal

3.1. Thermal stability

TGA and DTA patterns obtained in the temperature range of 300—670 K are shown in Figure 2. Rapid weight loss sets in at 500 K. It is therefore evident from the TGA

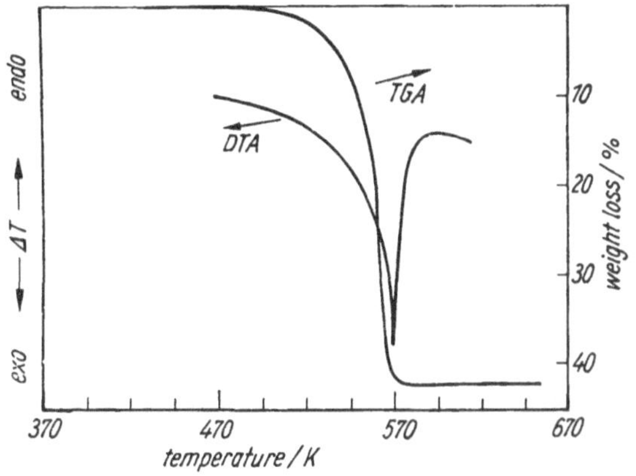

Fig. 2. TGA and DTA curves of KAP crystals

plot that KAP starts to decompose around 500 K at ambient pressure. The decomposition occurs over a narrow range of temperature (~ 60 K) and it is essentially completed by 568 K.

One would expect the products of the decomposition to be phthalic anhydride and potassium hydroxide according to the reaction.

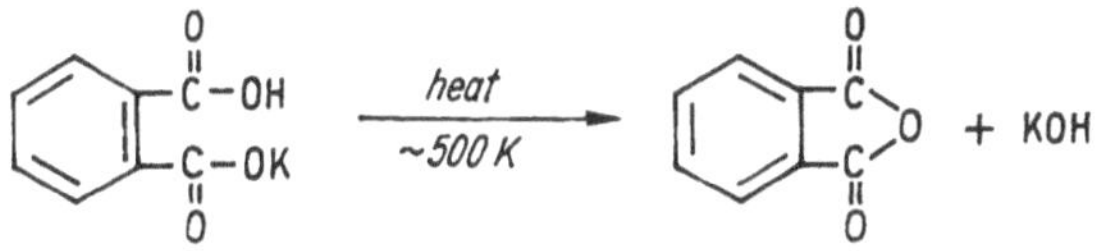

However, it was felt desirable to obtain independent experimental evidence to support this conclusion. Hence, powdered sample of KAP was heated to 570 K in a separate setup and the decomposition product, part of which sublimed was collected on the inner surface of a funnel kept enclosing the sample containing crucible. The sublimate was examined using X-ray diffraction and infrared spectra. The X-ray powder diffraction data of the sublimate was found to agree well with that of standard (sublimed) phthalic anhydride reported in the literature. The infrared transmission spectrum of sublimed product is compared with that of a standard sample at room temperature in Figure 3. It is confirmed that the sublimed decomposition product is phthalic anhydride. The residue left over in the container after sublimation was examined by standard analytical methods and it was confirmed to be KOH.

3.2. Dielectric studies

The dielectric constant of KAP single crystal was determined along the c-axis, as a function of both temperature and frequency. Dielectric constants were also determined along the three crystallographic axes at 1 MHz using Boonton capacitance meter. The values were found to be 6.2, 3.8, and 4.3 along a, b, and c axes, respectively, and are in good agreement with those reported in the literature (BELYAEV et al.). The dielectric constant measurements along the c-axis as a function of temperature and frequency are shown in Figures 4 (ε_r vs T) and 5 (ε_r vs $\log f$). Dielectric constant exhibits a quasi-linear rise with temperature in the range 300 to 420 K. This rise is consistent with the effect of temperature on the crystal structure of KAP. The structure consists of twin-layer blocks built of KAP units and the layers are coupled ionically. Within each layer of the block, KAP units are strongly bound by intermolecular hydrogen bonds. The blocks themselves are held by somewhat weaker van der Waals bonds which are responsible for the mica-like structure of the crystals; ultra-thin layers of KAP can

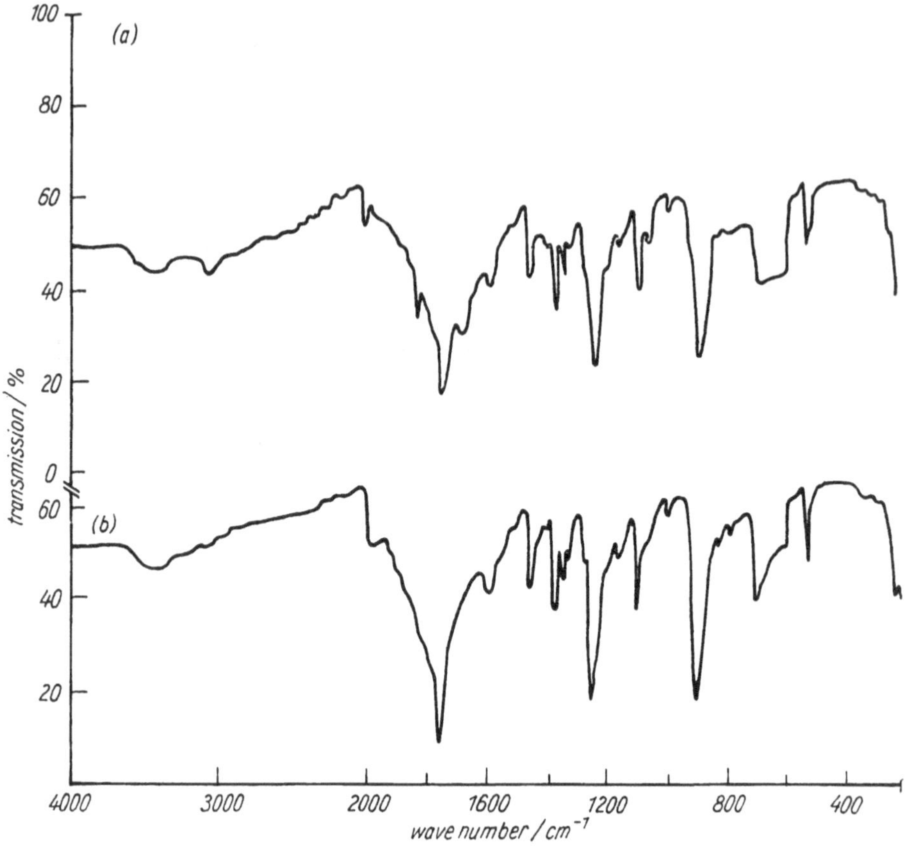

Fig. 3. Infrared transmission spectra of (a) standard (commercial Analar grade) phthalic anhydride and (b) sublimed phthalic anhydride recorded at 300 K

be peeled off perpendicular to the c-axis. The dipoles of the ionic $K^+ \dots 0^-$ pairs and the dipoles from the hydrogen bonded COOH groups are largely confined to ab plane in the middle of the blocks. The dipole moment along the c-axis is caused by the projection of these dipoles onto the c-axis. The measured dielectric constant along the c-axis is largely due to the contribution of these dipoles. Since the interblock bonding in the c-direction is van der Waals in nature, it is weak and is associated with large anharmonicity and hence large value of expansivity coefficient. When KAP is heated, due to thermal expansion the dipoles lying in the interlayer region also get stretched preferentially in c-direction, increasing the magnitude of the projections on to the c-axis. Thus the effective dielectric constant along the c-axis also increases.

The dielectric loss measured as $\tan \delta$ along the c-direction is plotted in Figure 4 for the same temperature range at 1 kHz frequency. Since the hydrogen bonds are broken at high temperatures it entails disordering effect upon the dipoles originating from COOH groups. It is likely to introduce disordering of the potassium ions also into the available equivalent sites. The net effect on the dielectric response of the disordering events is to increase the dielectric loss at a given frequency as is observed in Figure 4. The loss increases steeply above 400 K which suggests that a nucleation process preceeds the actual decomposition which starts around 400 to 420 K. The decomposition, however, becomes very rapid close to 510 K. The process of nucleation and the H-

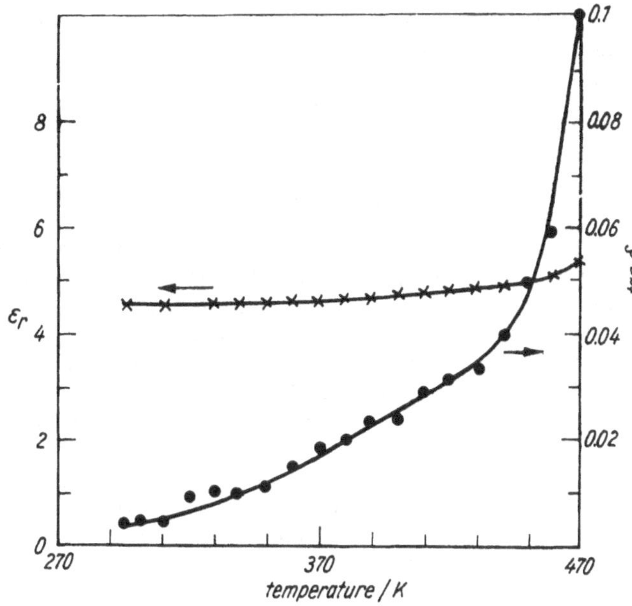

Fig. 4. Dielectric constant and tan δ as a function of temperature measured for KAP crystal along c-axis at 1 kHz

bond breaking are consistent with each other because the decomposition reaction involves elimination of phthalic anhydride intramolecularly leaving behind KOH. The proton in the OH group is actually involved in intermolecular H-bonding in the predecomposition stage. The nucleation therefore involves several steps such as breaking of the hydrogen bonds, reorientation of the OH groups and even repositioning of the K^+ ions in order to accomplish the decomposition. It is therefore initially governed by slow kinetics.

Dielectric constant of KAP crystal (Fig. 5) indicates that it is essentially frequency independent, however, the dielectric loss (Fig. 5) shows a small increase towards

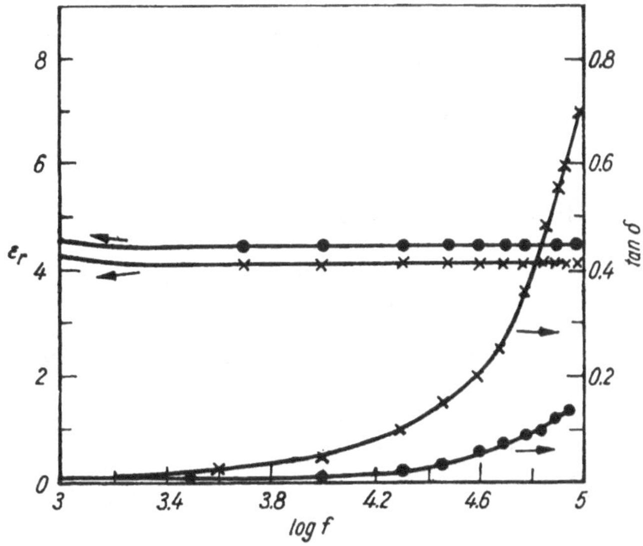

Fig. 5. Dielectric constants of X-ray irradiated (×) and unirradiated (•) KAP crystals and tan δ (parallel to c-axis) as a function of frequency at 300 K

higher frequencies. Since tan δ is equal to $(\varepsilon''/\varepsilon')$ and $\varepsilon_r = (\varepsilon'^2 + \varepsilon''^2)^{1/2}$, it implies that ε'' increases and ε' decreases (both slightly) at higher frequencies in the range of our measurement. In the light of the arguments presented previously, it appears that the small divergence in tan δ at high frequency end of the present study is largely a response of broken hydrogen bonds in equilibrium at laboratory temperature.

3.3. AC conductivity

The variation of ac conductivity of KAP is presented in Figure 6a as a function of temperature in the conventional Arrhenius plot and as a function of log f (frequency) in Figure 6b. At the outset, it is interesting to note that the 1 kHz ac conductivity which is sufficiently low is not as low as would be expected of a good insulator. Secondly, the ac conductivity does not vary significantly with temperature. In fact, the activation energy derived from the Arrhenius plot is equal to 0.022 eV which is low even for typical hopping barrier. Since the phthalate consists of hydrogen bonds it is tempting to associate the electrical transport in KAP with protonic conductivity. These protons are likely to tunnel across the layered blocks accounting for conductivity along c-axis. The conductivity is expectedly low since only a few protons can be expected to be involved in the transport process. The barrier for this tunneling process is also understandably very low.

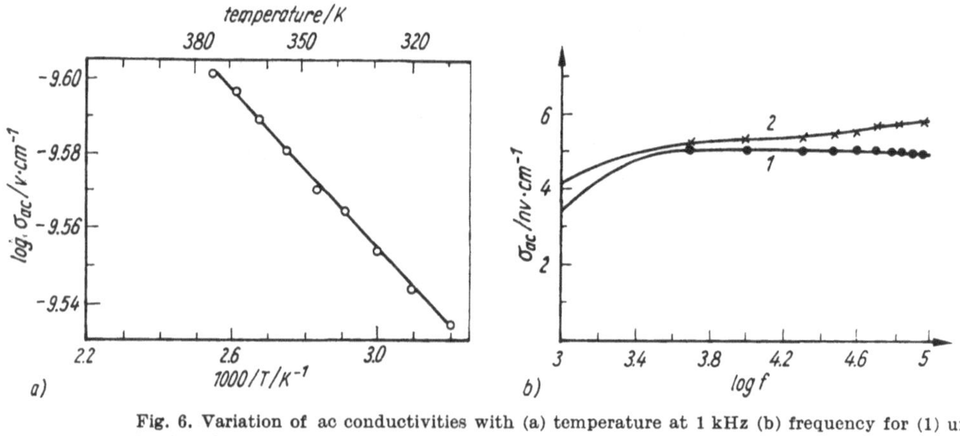

Fig. 6. Variation of ac conductivities with (a) temperature at 1 kHz (b) frequency for (1) unirradiated and (2) X-ray irradiated (300 K) crystals of KAP along c-axis

The frequency dependence of the conductivity is also very low (Fig. 6b) and is insignificant particularly in the high frequency region between 10 to 100 kHz. However, in the region upto 10 kHz conductivity appears to increase as is normally expected. From Figures 4 and 5, it may be noted that above 10 kHz, tan δ increases quite pronouncedly although ε_r remains constant. Since $\sigma_{ac} = \omega\varepsilon$ tan δ the increase in σ_{ac} with ω should be compensated by the product ε' tan δ. Further since tan δ only increases slightly in this region ε' should decrease which is consistent with the results of earlier section. It is, however, unclear as to why the real part of the dielectric constant decreases with increase in frequency. It is thoughtful to note in this context that the single crystal itself may be viewed as a stacking of parallel plate condensors in the c-direction and frequency response must be a function of thickness of the crystal. This aspect is being further investigated.

3.4. Effect of X-ray irradiation

The effect of X-ray irradiation of a variety of dicarboxylic acids has been discussed in literature. The primary event appears to be the production of free radicals by breaking OH bond of acid groups. In the case of KAP, there are many other candidate centers for the formation of free radicals. It was our interest to see whether the cleavage of such bonds would give rise to functional group migrations which might eventually eliminate the two dimensional character of KAP.

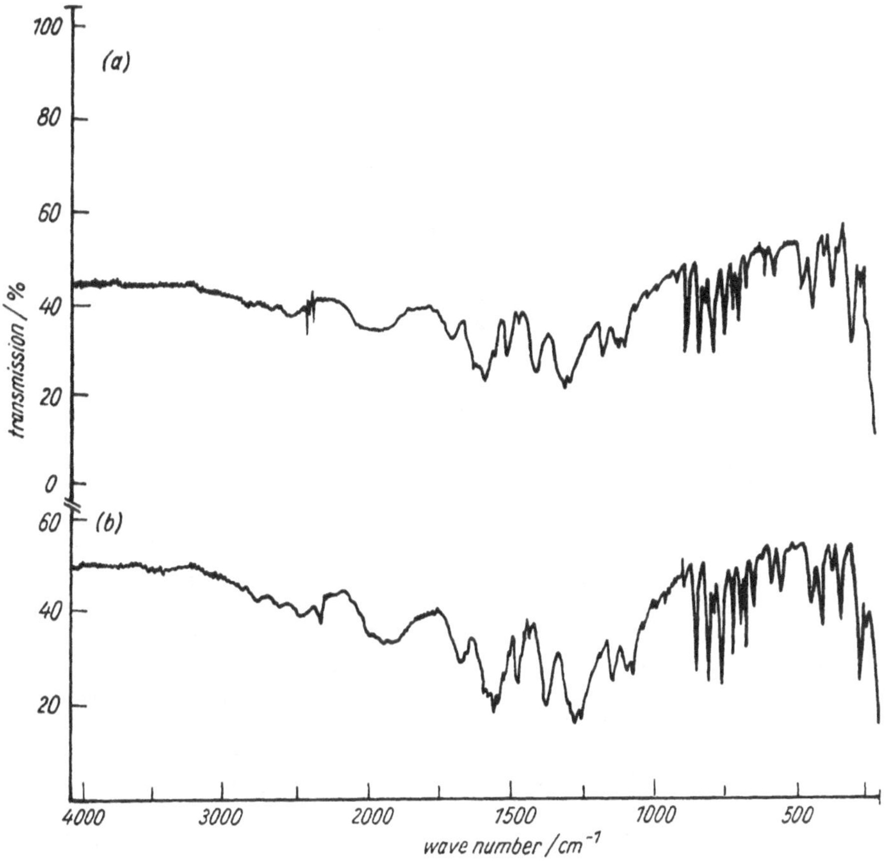

Fig. 7. Infrared transmission spectra of (a) unirradiated and (b) X-ray irradiated crystals of KAP at 300 K

KAP single crystal was heavily irradiated as described in the experimental section The crystal turned into pale yellow colour which persisted indefinitely. The dielectric and transport properties of irradiated crystals were studied alongside those of unirradiated samples. The results of dielectric and ac conductivity studies performed as a function of frequency on the irradiated crystals are presented in Figures 5 and 6 (see Figure legends).

The relative dielectric constant slightly decreases in the entire frequency range. Dielectric loss increases as a function of frequency more prominently towards higher frequencies as compared to unirradiated samples (Fig. 5). The irradiated crystal of KAP was annealed at 400 K for 10 hours and gradually cooled to room temperature. The dielectric constants were measured at 1 MHz for the cooled sample and the dielectric constants were found to be restored to their original values.

The infrared spectra of the irradiated and unirradiated samples recorded in KBr matrix is presented in Figure 7. The infrared spectra clearly indicate that no features

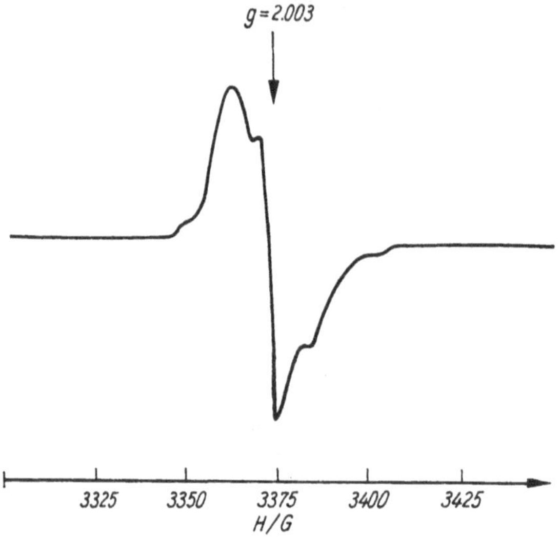

Fig. 8. ESR spectrum of X-ray irradiated single crystal of KAP with the magnetic field parallel to c-axis at 300 K

attributable to any structural modification are present. The irradiated coloured samples gave rise to ESR signals around $g = 2.003$ at 300 K which is shown in Figure 8. It is clear from ESR spectrum that the free radical is stabilized and the $H^{\cdot}$ attaches itself to a suitable position most probably to ring carbon atom. It is thus possible that there are two spin centers which give rise to the observed ESR signal. In principle, electron on the carbon atom of the ring should produce fine structure consisting of six lines which in the present 300 K spectra is not evident. The irradiation has the same effect as breaking of the hydrogen bonds in thermal experiments but without the presence of the associated thermal expansivity. This appears to suppress the dipole-moment contribution of the COOH groups in the c-direction, accounting for the apparent decrease in dielectric constant. The pronounced dielectric loss at higher frequencies is quite consistent with increased disorder due to free radical production. Since the free radical production may revert partly to ion pair production, we would expect the protonic conductivity to increase slightly and this is evidenced in Figure 6b.

The authors are thankful to Prof. C. N. R. Rao for his kind encouragement and Mr. S. Anantharaj for helpful discussions.

References

BELYAEV, L. M., BELIKOVA, G. S., GIL'VARG, A. B., SIL'VESTROVA, I. M.: Sov. Phys.-Crystallogr. **14** (1970) 544

CHUMAKOV, A. A., KOPTSIK, V. A.: Sov. Phys.-Crystallogr. **4** (1960) 212

DISTLER, G. I., SUVOROVA, E. I., GERASIMOV, YU. M., KORTUKOVA, E. I., OKHRIMENKO, T. M., BELIKOVA, G. S.: Sov. Phys. Solid State **27** (1985) 281
GAVRILOVA, N. D.: Sov. Phys.-Crystallogr **10** (1965) 91
GROTH, P.: Chemische Kristallographie **4** (1906) 724
JONES, J. L., PASCHEN, K. W., NICHOLSON, J. B.: J. Appl. Optics **2** (1963) 955
JETTEN, L. M. A. J., VANDERHOEK, B., van ENCKEVORT, W. J. P.: J. Cryst. Growth **62** (1983) 603
KOPTSIK, V. A., MINAEVA, K. A., VORONKOV, A. A., SOLOV'EV, A. F., IZRAILENKO, A. P., POPKOVA, E. G., KOZLOVA, G. I.: Vestn. Moscow. Univ. **61** (1958) 6
VAN ENCKEVORT, W. J. P., JETTEN, L. A. M.: J. Cryst. Growth **60** (1982) 275

(Received May 27, 1987)

Authors' address:
c/o Prof. K. J. RAO
Materials Research Laboratory, Indian Institute of Science
Bangalore 560012, India

Book Review

G. E. Dieter

Mechanical Metallurgy

3rd ed., Mc Graw-Hill Book Co., New York 1986. XXIII + 751 p., DM 138.50. ISBN 0-07-016893-8

The subject of this book is the response of metals to applied mechanical forces and its objective has been to include the entire scope of that subject in one comprehensive volume. Emphasis is on basic phenomena and on engineering aspects. The materials under considera-tion are considered both from a continuum and an atomistic point of view. There is some tendency to make this book a handbook covering all relevant topics and providing nu-merous references to the literature. In addition, a bibliography of general references can be found at the end of each chapter. Most topics are discussed in greater detail than when they were presented in an undergraduate course. In fact the book is written for graduate students and practicing engineers who need an up-to date overview as well as an introduction to the fundamentals.

The volume is divided into four parts. Part one and two deal with the mechanical and metallurgical fundamentals necessary for a comprehension of what follows. They cover an introduction to stresses and strains in elasticity and plasticity and to the more physical aspects of strengthening and fracture. Part 3 gives a survey of the common testing tech-niques, i.e. tension, hardness, torsion, fracture, fatigue and creep. Emphasis is placed on the interpretation rather than on the procedures. The final part deals with the common mechanical processes for producing useful metal shapes such as forging, rolling, extrusion, drawing of rods a.s.o. and machining. Numerous illustrations and selected problems (incl. answers) have been included emphasising the text book character.

Both SI and American engineering units are used with emphasis on SI units. The present edition is again subject to substantial extension. A separate chapter on fracture mechanics, deformation maps, finite element methods and the creep-fatigue interaction have been added among other catch words.

The main advantage of this book is related to the fact that a single author has covered such a wide field. It is well-produced and should be accessible to all students of engineering disciplines.

P. Paupler

Cryst. Res. Technol. | 23 | 1988 | 2 | 195–198

C. C. Desai

Department of Physics, Sardar Patel University Vallabh Vidyanagar, Gujarat, India

Recent Advances in the Theory of Crystal Growth in Silica Gel Medium

Growth of single crystal materials by slow diffusion process in a silica-gel medium has been used for a long time as the method of simulation of natural processes in the laboratory. Recent advances in the theory of crystal growth have been described. The suitable theory of nucleation has been developed. Attention has been paid mainly to the more general case in which the counter diffusion of two reacting materials occurs. The implications are discussed.

1. Introduction

Gel media appear to be the best for the controlled growth of a variety of crystalline materials suitable for research and technology. The method is specifically useful for substances which, because of their low solubilities, or low dissociation temperatures, cannot readily be grown in other ways. The gel medium prevents turbulence and remaining chemically inert, provides a three-dimensional structure, which permits the reagents to diffuse at a desirable controlled rate. Further, its softness and the uniform nature of constraining forces that it exerts upon the growing crystals encourages orderly growth. The growth of crystals is affected by various parameters such as gel pH, gel ageing, gel quality, gel density, environmental temperature, nature and concentration of acids, purity and strength of reactants (supernatants), type of crystallization apparatus, etc. But the processes working in the growth of good quality, transparent and perfect crystals in the system remain ignored. The aim of this paper is to report on recent advances in the theoretical aspect of crystal growth in silica gel medium.

2. Theory of nucleation and growth mechanism

Studies dealing with the theoretical aspects of the crystal growth in silica gel medium has been reported by various investigators (Kirov; Srzic et al.; Desai). In all the cases, main attention is paid to the more general case in which the counter diffusion of two reacting substances occurs. It is assumed that in low temperature diffusion method, each one of the reactants travels through the porous medium of silica gel according to Fick's law:

$$D\left[\frac{\partial^2 c}{\partial x^2}+\frac{\partial^2 c}{\partial y^2}+\frac{\partial^2 c}{\partial z^2}\right]=\frac{\partial c}{\partial t} \tag{2.1}$$

where

C concentration of reacting solutions (reactants)
D coefficient of diffusion
t time of diffusion.

Here, the following assumptions are to be considered:

(i) the diffusion coefficient (D) is a constant, and therefore it is independent on the concentration of the reactants, and

(ii) only linear diffusion is considered. Therefore equation (2.1) is reduced to:

$$D\left[\frac{\partial^2 c}{\partial x^2}\right] = \frac{\partial c}{\partial t} \tag{2.2}$$

Now, each of the reactants slowly diffuses into a semi-infinite elongated gel column and is diagrammatically shown in Figure 1. Therefore, the initial and limited conditions are the following:

$$\begin{aligned} &\text{at } x = 0\,, \quad c = c_0\,, \quad t > 0 \\ &\text{at } t = 0\,, \quad c = c_0\,, \quad x > 0 \end{aligned} \tag{2.3}$$

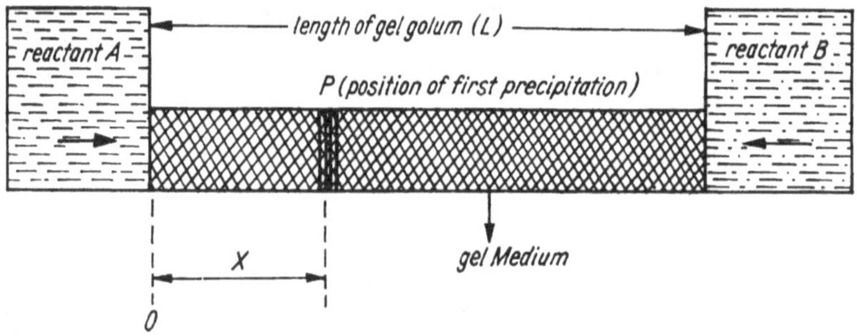

Fig. 1. A diagram of the U-tube arrangement. C_{OA} and C_{OB} show initial concentration of reactants A and B. P is the position of the first precipitation

In these conditions the partial differential equation (2.2) has the following solution:

$$C = C_0 \operatorname{erf}\left[\frac{x}{2\sqrt{D \cdot t}}\right] \tag{2.3}$$

where erf = Gauss error function complement.

For the reacting solvents A and B (Fig. 1), the concentration variations will be given by:

$$C_A(x, t) = C_{OA} \operatorname{erf}\left[\frac{x}{2\sqrt{D_A \cdot t}}\right], \tag{2.4}$$

$$C_B(x, t) = C_{OB} \operatorname{erf}\left[\frac{L - x}{2\sqrt{D_B \cdot t}}\right] \tag{2.5}$$

where

L length of the gel column
D_A diffusion coefficient for reactant A
D_B diffusion coefficient for reactant B.

The product of A and B concentrations will be given by

$$C_{AB}(x, t) = C_A(x, t) \cdot C_B(x, t) \tag{2.6}$$

Substituting equations (2.4) and (2.5) in equation (2.6), we have

$$C_{AB} = C_{OA} \cdot C_{OB} \operatorname{erf}\left[\frac{x}{2\sqrt{D_A \cdot t}}\right] \cdot \operatorname{erf}\left[\frac{L - x}{2\sqrt{D_B \cdot t}}\right] \tag{2.7}$$

The first precipitation will be formed when the solubility product of the crystallizing compounds is reached, i.e. when equation (2.7) presents a maximum value. This maximum value does depend on the initial concentrations of the reactants. However, this assertion disagrees with the experimental observations. It is well demonstrated that for the same crystallizing compound, the position of the first precipitation varies with the ratio of initial concentrations of A and B reactants. This illustrates that the classical law of the precipitation theory cannot be applied to the case of low temperature diffusion method.

An alternative model has been proposed. This model emerges by considering the formation of the first precipitation as predicted by the classical nucleation theory.

The cluster of i^* atom, the chemical potential of which is in equilibrium with that of the metastable phase, is called "critical nucleus". Any cluster of i atoms such that $i > i^*$, tends to grow. On the contrary, a cluster of i atoms being $i \cdot i^*$ tends to dissolve. Therefore, one can consider such a situation where the solubility product of the substance is reached, but stoichiometric crystal nuclei cannot be initiated. For example, let us consider the case of a solution with 2 atoms A and 2 atoms of B type. After transformation of atomic concentrations to molarities, the solubility product AB is reached. However, larger nuclei, than those containing 2A + 2B atoms cannot be formed and thereby tends to dissolve.

Once the solubility law, i.e. $C_A \cdot C_B > K_s$, is established, it is required that the smaller equivalent concentration equals the square root of a critical value $K_S^* > K_S$. The difference between K_S^* and K_S is a result of the free energy increment for the formation of the critical size nuclei. Above situation can be imagined where:

$$\begin{aligned} C_A &\neq C_B \\ C_A \cdot C_B &= K_S^* . \end{aligned} \tag{2.8}$$

In the case of low temperature diffusion method, it can be illustrated that the above considered requisite for the formation of nuclei is initially obtained when:

$$C_A = C_B = \sqrt{K_S^*} . \tag{2.9}$$

This equation controls the formation of first precipitation when counter-diffusion of the two reacting solutions occur.

Desai, Rai, and Hanchinal have shown agreement between theoretical results and experimental observations. They proposed a model to predict the position of first precipitation in relation to the critical concentration of the reactants and to the values of diffusion coefficients and the width of the precipitation zone. The concentration rate of the reactants in the region where the first precipitation will appear, is given by:

$$C_A \cdot C_B = C_{AB} = C_{0A}^2 \operatorname{erf}^2 \left[\frac{1}{2\sqrt{t}(\sqrt{D_A} + \sqrt{D_B})} \right] . \tag{2.10}$$

For the case of $C_A \cdot C_B = K_S^*$, the following three different zones in the crystal growth can be considered:

(i) The precipitation occurs at very high supersaturation as a sudden even. Nucleation occurs in the non-classical range (Sears; Stoyanov) and therefore "amorphous precipitation" will be formed.

(ii) The intermediate region belongs to a softening of the concentration rate and thereby "spherulites" (polycrystalline aggregates) will be formed.

(iii) Substances with K_S^* higher than 10^{-6}, thereby nucleation initiates, where equilibrium is slowly reached and therefore single crystal can be generated. In order to check the results, critical concentration being unknown, the related values K_S^* must be used upto now.

Following another concept (HENISCH) supposes that the solute supersaturation φ_∞ at large distances from the crystal remains unchanged during crystal growth. According with Frank theory

$$\varphi_\infty - \varphi_0 = f\left(\frac{R}{\sqrt{Dt}}\right) \tag{2.11}$$

where

φ_0 solute supersaturation just at the crystal surface
R radius of the crystal, assuming it is spherical
D coefficient of diffusion
t time of diffusion.

It is observed that a constant value of $\left(\frac{R}{\sqrt{Dt}}\right)$ implies a constant value of φ_0. In this case the following relation must be accomplished:

$$R^2 = K + Dt \tag{2.12}$$

which has been many times verified, suggesting thereby that the chemical reaction was not the rate-determining process in crystal growth and the existence of a constant supersaturation in the crystal surface. However, it is important to note that some deviations from relation (2.12) have been found for extreme t values. This suggests different growth mechanism at earlier and later stages of the crystal growth.

The equation (2.12) which is typical of one-dimensional diffusion process, has been experimentally verified, for instance, among others by BLANK, by LIAW and FAUST, and by DESAI and RAI, and DESAI and HANCHINAL. This theory allows us to predict a future development of the laboratory simulation of natural (geological, and biological) crystallising systems.

The author is thankful to University Grants Commission (New Delhi) for the sanction of the project entitled "Studies of growth, characterization and properties of alkali halides and tartrate single crystals". CCD is very grateful to Mrs. Parul, Bhavina, and Mr. Ramesh Desai (Union, N.J. 07083, USA) for their interest and encouragement.

References

BLANK, Z.: J. Cryst. Growth **18** (1973) 281
DESAI, C. C.: Cryst. Res. Technol. **22** (1987) 1117
DESAI, C. C., RAI, J. L.: Surface Technology **15** (1981) 15
DESAI, C. C., HANCHINAL, A. N.: Surfacce Technology **25** (1985) 69
HENISCH, H. K.: Crystal Growth in Gels, Pennsylvania 1970
KIROV, G. K.: J. Cryst. Growth **15** (1972) 102
KIROV, G. K.: Compt. Rend. Acad. Bulg. **30** (1977) 559
LIAW, H. M., FAUST, J. W.: J. Cryst. Growth **8** (1971) 8
SEARS, G. W.: J. Chem. Phys. **65** (1961) 1738
SRZIC, D., POKRIC, B., PUCAR, Z.: Z. Physik. Chem. **103** (1976) 157
STOYANOV, N.: Materials Science, Vol. 3, Edited by E. KALDIS, Amsterdam 1979

(Received June 17, 1987)

Author's address:
Professor C. C. DESAI
Department of Physics, Sardar Patel University
Vallabh Vidyanagar 388120
Gujarat, India

Cryst. Res. Technol.	23	1988	2	199—207

M. Suszyńska, R. Capelletti

Institute of Low Temperature and Structure Research, Polish Academy of Sciences, Wrocław, Poland, and
Istituto di Fisica dell 'Universita, Parma, Italy

Thermally Stimulated Depolarization Currents in OH^--doped NaCl and KCl Crystals

The main features of the high-temperature ITC band are analyzed on the basis of experimental results obtained for NaCl and KCl crystals containing the hydroxyl ions. The unusual behaviour of this band has been compared with that described previously for alkali halides, AH, (LiF, NaCl, KCl) doped with some divalent cations, Me^{++}, (Be, Eu, Ni, Pb).
It has been shown that — similar to the case of the Me^{++}-doping-most of the results obtained can be explained in frames of the Maxwell-Wagner-Sillars model of interfacial polarization induced by the impurities forming the charge clouds to the dislocation lines. An ITC band can be detected due to the fact that these regions exhibit electric conductivity and dielectric constants different from those characteristic of the bulk material. The role of the charge of dislocations in the induced polarization phenomenon is shortly discussed.

1. Introduction

The ionic thermo-current, ITC, technique, generally known as the thermally stimulated depolarization current (TSDC) method, is mostly exploited for the detection of relaxation phenomena related with the cation impurity — cation vacancy (I.V.) dipoles, and hence limited to the low-temperature depolarization range in the case of alkali halides (AH) doped with divalent impurities (Me^{++}) (Bucci et al. 1964, 1966). But, in principle, this method can be applied for the detection of any kind of thermally activated dielectric process in solids exhibiting negligible electronic conductivity.

As has been shown previously (Capelletti et al. 1976, 1985; Suszyńska et al. 1978, 1979, 1984, 1985), the ITC spectra of the $AH:Me^{++}$ systems, measured in an extended temperature range (up to about 400 K), exhibit the presence of a huge signal, hereafter called the C-band. Some of the main features of this band are unusual in comparison with features of the well known depolarization peak related with the presence of I.V. dipoles. The origin of this band has been tentatively attributed to a Maxwell-Wagner type relaxation induced by the Cottrell-like atmospheres distributed around the dislocation lines. This hypothesis obtained a direct support from the experiments with plastic deformation and from the recovery of deformation-induced effects. Some of the results obtained could be also explained by an alternative model concerning the bulk conductivity partly blocked by the insulating air-gaps occurring between the sample and the measuring electrodes (Müller).

In order to obtain further experimental evidences supporting one of the mentioned explanations, systematic studies of the C-band were undertaken for alkali halides containing the hydroxyl ions (OH^-) either as dopant- or as impurity-ions. It should be noted that in contrast to the $AH:Me^{++}$ systems the charge on the dislocation lines present in these crystals is found to be positive (Davidge; Whitworth).

2. Experimental details

2.1. Samples origin and preparation

The specimens of monocrystalline sodium and potassium chloride, containing the hydroxyl anions as the dopant and/or as impurities were cleaved from as grown crystal ingots of various origin. The list of the crystals investigated as well as some informations concerning their origin, doping, and/or contamination level, are given in Table 1.

Table 1
Some characteristics of the crystals

Material		impurity content		remarks
mark	origin	c_{OH} in the melt mppm	α cm^{-1}	
NaCl : OH^-				
22	Harshaw	n.p.	0.31	1. Super pure NaCl and KCl crystals, used for the effects of irradiation, have been grown by the modified Bridgman method in our Lab. in Wrocław
15 }	Crystal Growth	100	11.46	
16 }	Lab., Budapest	500	75.00	
KCl : OH^-				
27 }	Karl Korth Lab.	100	optically not	2. Crystals from Harshaw and Hilger Watt are approximately 30 years old, and have been sold as nominally pure ones
28 }	Kiel	1000	detectable	
21	Harshaw	n.p.	0.65	
20	Hilger Watt	n.p.	0.80	
19 }	Crystal Growth	50	1.90	3. Crystals 25 and 26 were grown by the Kyropoulos method in dry nitrogene atmosphere
18 }	Lab., Budapest	500	68.00	
25 }	Physics Depart-	unknown	8.10	
26 }	ment, Parma		51.70	

The actual concentration of OH^- in KCl and NaCl crystals, c_{OH}, was determined chemically by the titration of water solutions of the samples acc. to the methods described by FRITZ et al., and SUSZYŃSKA et al. (1982), respectively. These data were compared with those obtained by measuring the optical absorption coefficient of the appropriate absorption bands positioned in the ultraviolet range, α_{OH}, i.e. at 185 nm for NaCl (KESSLER) and at 204 nm for KCl (FRITZ et al.). The slope of the linear (α_{OH} [cm^{-1}] vs c_{OH} [mole ppm])-dependence was equal to 0.342 and 0.25 cm^{-1} per 1 mole ppm of OH^- for NaCl and KCl crystals, respectively, The reproducibility of the chemical determination of c_{OH} was equal to ± 3 mole ppm for samples taken from the same place of the ingot.

Measurements have been chiefly performed for as grown samples. In order to check the effect of dislocations, some of the samples were plastically deformed. The recovery of the deformation-induced effects was studied at room temperature (RT) for one representative of each type of crystal.

For the purposes of this work two sample geometries were used:

1. plates of about 100 mm^2 cross section, and the thickness ranging between 0.3 and 2.0 mm; the thickest samples were usually used for the optical absorption measurements, and
2. parallelepipeds of about $3 \times 3 \times 9$ mm^3 for plastic deformation performed under conditions of uniaxial compression along a $\langle 001 \rangle$ direction; after deformation the ITC was monitored across the less changed sample thichness.

2.2. Measurement techniques

The measurements of thermally stimulated depolarization currents were performed by using the ITC — set up. For general experimental details concerning this measurement technique see the works by BUCCI et al. (1964, 1966). For the purposes of this work the measurements were performed in the temperature range between 70 and 400 K, using a vibrating reed electrometer. The detection limit of the thermally stimulated current was equal to about 10^{-16} A.

The optical absorption spectra for the plate-shaped samples used for ITC measurements have been controlled by using the VARIAN 2390 and JASCO 702 G spectrophotometers for the measurements in the UV- and IR-spectral ranges, respectively. For the purposes of the α_{OH}-determination a BEAUDUIN-MVR-100 vacuum monochromator with a windowless hydrogen lamp was additionally used. The IR-measurements were performed by using a double-beam BECKMAN-DK-ZA-spectrophotometer. The registration of the transmission spectra was performed by using a chart-scale of 10 nm per 1 cm, and a chart-speed of 3,6 cm per minute.

3. Results and discussion

3.1. Features of band-C in crystals with OH^-

Figure 1 presents three high-temperature parts of the ITC spectra obtained for a KCl crystal containing approximately 2 mole ppm of OH^-. Curve "a" of this figure corresponds to a sample prepared from the as grown crystal, curve "b" shows the effect of plastic deformation, and curve "c" represents the result of annealing performed at room temperature.

The main effect of plastic deformation is to suppress the amplitude of band C (H^c) and to shift the position of its maximum (T_M^c) towards lower temperatures. The degree of RT-recovery of the effect induced by plastic deformation depends on the OH^- concentration, being more complete for more dilute samples. Also the changes of T_M^c are concentration dependent, whereby the value attained is for the most dilute samples larger than the T_M^ccharacteristic of the undeformed specimens. Moreover, the shape of the recovering C-band, and hence the activation energy of the process studied, differs from that characteristic of the as grown sample, whereby both crystal systems behave in a different way. Whilst for NaCl:OH^- crystals the recovering C-band is broader on the low-temperature side, for KCl:OH^- crystals the reverse behaviour was registrated. It should be emphasized that the recovery process was studied at RT for a maximum annealing time equal to 24 h. Hence, this experiment cannot confirm if the crystal state produced by plastic deformation can be or cannot be transformed completely to the initial state by appropriate thermal treatment.

Figure 2 shows the relations between the polarizing voltage (V_p) applied to various specimens of KCl crystals and their current response (H^c). The linearity of these relations suggests that the band considered refers to a bulk polarization process. In contrast to the $H^c(V_p)$-dependence, the position of the maximum of band C, T_M^c, is independent on the electric field applied (for experiments performed under conditions of constant sample thickness). Qualitatively similar results have been obtained for all other crystals. Besides the effect of V_p upon T_M^c and H^c the dependence of both parameters on the polarization time (t_p) and polarization temperature (T_p) was studied. It has been found that for a typical values of $t_p \geqq 3$ min the values of T_M^c were independent on the polarization time. The value of T_M^c was also independent on T_p, whereas H^c is an increasing function of T_p for the experimental conditions $T_p < T_M^c$.

Figures 3a and 3b show the concentration dependences of H^c and T_M^c, respectively, for both crystal systems investigated. By plotting these dependences of H^c vs c_{OH} in

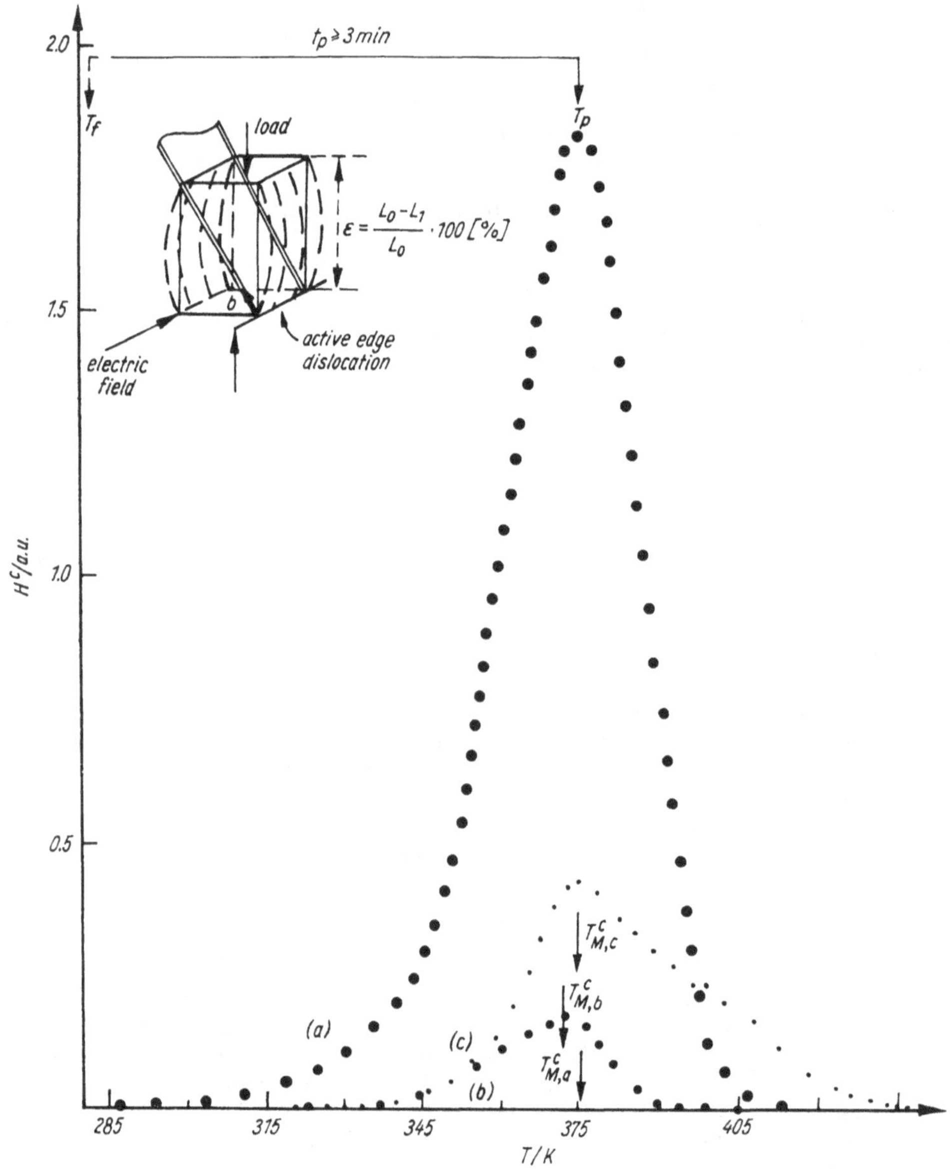

Fig. 1. C – band characteristic of the KCl crystal containing approximately two mole ppm of OH^-; curves a, b, and c represent the high-temperature band of as grown, plastically deformed, and recovered sample, respectively

a double logarithmic scale, no simple relations have been found for the concentration range studied, suggesting that the process responsible for the relaxation is not directly related with the contamination level. One possible reason for this — rather unexpected — result could be related with the fact that the crystals used in this study are of different origin, and different growth conditions yield different density of grown-in dislocations, disguising the effect of the OH^- ions studied. Another unexpected result concerns the $T^c_M(c_{OH})$-dependence obtained for the NaCl:OH^- system. Namely, yet

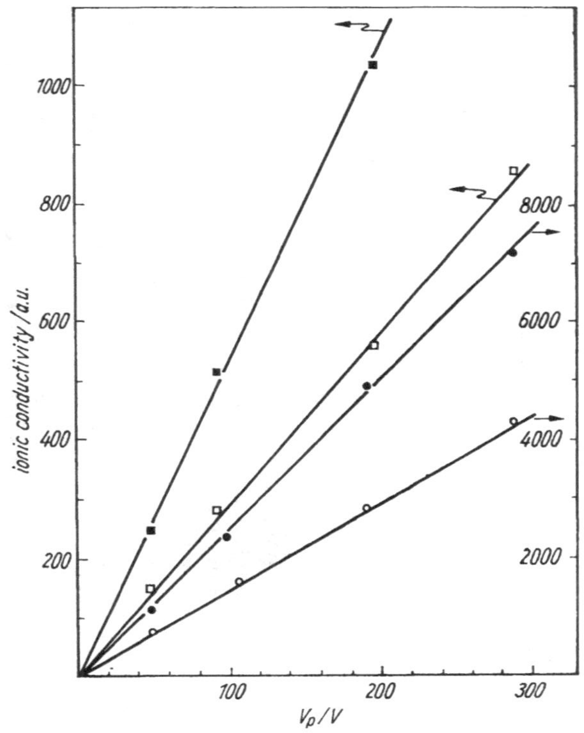

Fig. 2. Ionic conductivity versus polarization voltage for various KCl samples There are shown the effects of dopant concentration (o and ● for samples 27 and 28, respectively), sample thickness (■ and □ for x = 2.18 and 3.80 mm of a deformed sample 28), and plastic deformation (2.8 %)

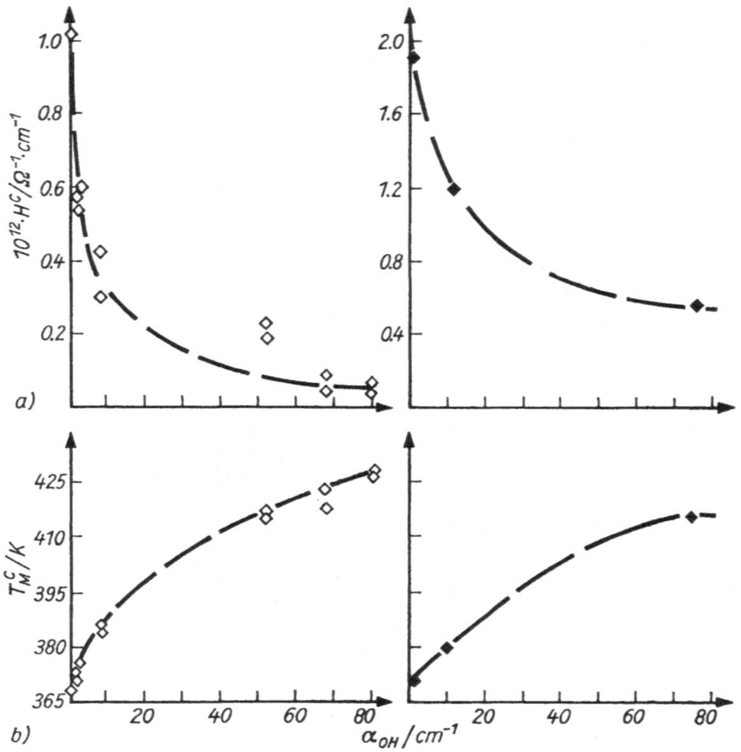

Fig. 3. Concentration dependences of the band pmplitude (Fig. a) and the band location (Fig. b) open and full marks are for the $KCl:OH^-$ and $NaCl:OH^-$ systems, respectively

for the diluted crystals the T_M^c-value is much higher than any T_M^c characteristic of nominally pure NaCl crystals (SUSZYŃSKA et al. 1984).

3.2. Comparison with the data characteristic of $AH:Me^{++}$

As has been shown previously (CAPELLETTI et al. 1976, 1985; SUSZYŃSKA et al. 1978, 1979, 1984, 1985), the high-temperature ITC band appearing in the $AH:Me^{++}$ systems is a single nonsymmetric one with the following features:

I. For as grown systems:

1. The position of band C is independent of the polarization conditions but it is influenced by the dopant concentration (c_{++}) and the sample thickness (x) as well.
2. The band amplitude is independent of $t_p \geqq 3$ min, and it depends on the sample thickness, polarization voltage (for x = const), polarization temperature (for $T_p < T_M^c$), density of grown-in dislocations, dopant concentration and the dopant dispersion state.
3. The amplitude of band C is approximately two orders of magnitude larger than that related with reorientation of I.V. dipoles.
4. The shape of this band, and hence the activation energy related with the relaxation process considered, is independent of the sample thickness.

II. The special role of dislocations surrounded either by the charge-compensating species (e.g. the I.V. dipoles in $KCl:Me^{++}$ crystals) or by precipitate particles (in the case of $NaCl:Me^{++}$ systems) has been evidenced by looking to the following effects:

1. The role of grown-in dislocations; in the case of $KCl:Me^{++}$ crystals the increasing dopant concentration is accompanied by the decreasing density of grown-in dislocations (ϱ).
2. The experiments with plastic deformation performed under controlled conditions of uniaxial compression; the decrease of H^c was proportional to the square root of the density of dislocations present in the deformed specimens.
3. Thermal recovery of the deformation-induced effects; the kinetics of this process follows a $t_{ann}^{0,66}$-law typical of preferential segregation of impurities in the stress fields of dislocations (COTTRELL et al.).

The typical behaviour of band C in the $AH:Me^{++}$- and $AH:OH^-$-crystals have been put together in Table 2. As can be seen, the main differences in the behaviour of this band relate to the following features of the band characteristics:

- the concentration dependence of H^c,
- the effect of deformation upon T_M^c, and
- the changes of T_M^c during the RT-storage of the deformed specimens.

3.3. Interpretation of the data

The dependency upon polarization conditions and the concentration — dependence of both parameters characterizing the C-band point out that in the OH^- containing alkali halides this band is due to a polarization phenomenon related with the impurities in a more or less definite way. The linearity of the $H^c(V_p)$-dependence as well as the independency of T_M^c on T_p and V_p show that band C does not relate to a space-charge phenomenon. The hugeness of this band speaks against dipolar point defects, whereas its nonsymmetrical shape — against diverse dipolar aggregates, cp. the work by CAPELLETTI et al. (1983).

Because of these facts and because of the similarities to the $AH:Me^{++}$ systems, it was assumed that the Maxwell-Wagner-Sillars model of interfacial polarization in

Table 2
Main features of band C in doped alkali halides

Function	Crystal system: Me^{++} in: KCl	Me^{++} in: NaCl	OH^- in: KCl	OH^- in: NaCl
effect of V_p (at x = const) on:				
T_M^C	← does not depend →			
H^C	← linearly — dependent →			
effect of T_p on:				
T_M^C	← is not influenced →			
H^C	← increases for $T^p < T_M^C$ →			
effect of c_{++}, c_{OH} on:				
T_M^C	decreases	increases	← increases →	
H^C	← increases →		← decreases →	
effect of deformation (ε) on:				
T_M^C	shifts towards higher values		shifts towards lower values	
H^C	← strongly decreases →			
recovery of ε for:				
T_M^C	← practically not recoverable →		concentration-dependent changes	
H^C	← concentration-dependent recovery →			

heterogeneous dielectrics could in principle also explain the effects reported in the present work. The dislocation lines together with the vacancies and/or impurities, which form the charge compensation clouds and/or decoration centers, respectively can be considered as inhomogeneities the electrical conductivity and dielectric constant of which differ from those characteristic of the crystal matrix. For some quantitative considerations cp. the work by CAPELLETTI et al. (1985).

The most convincing experiment which points at the role of dislocations relates to the effect of plastic deformation which affects both the size of the atmospheres and the density of dislocations as well. Taking into account that the charge of dislocations in AH:OH^- crystals is opposite to that characteristic of the AH:Me^{++} systems, the deformation induced changes of T_M^c are explainable by the polarization phenomena rather than by the bulk conductivity model suggested by MÜLLER. Unfortunately, in contrast to the AH:Me^{++} systems no serious experimental data exist which concern the deformation induced changes of ionic conductivity of the AH:OH^- crystals. Taking again into account the opposite dislocation charge in both systems, and assuming that cation vacancies are mainly responsible for the ionic conductivity, one could expect that the deformation of AH:OH^- crystals should yield a decrease of conductivity. In frames of the MÜLLER's model the decrease of conductivity should give a decrease of H^c and an increase of the T_M^c-value, not confirmed experimentally, cp. Sect. 3.1.

It should be emphasized that the problem concerning the origin of the positive charge of dislocations in OH^- doped alkali halides is stil an open one. For OH^- entering the "monovalent" ionic lattice substitutionally, no charge compensation is necessary. It is suggested, however, (FRITZ et al.; KESSLER) that such crystals always contain some quantities of CO_3^{2-}- and/or O^{2-}-ions which need the introduction of an equivalent number of anion vacancies the effective charge of which is positive. These vacancies may be swept-up by the dislocations being responsible for their positive charge.

Another possibility exists for the positive charge of dislocations in AH : OH^- crystals if one assumes the presence of some divalent cation impurities which react with the OH^--ions. Acc. to the four-site-model proposed by Fritz et al. for ($KCl + Ca^{++} + OH^-$) crystals, the formation of an $[(OH)_2]^-$ molecule-ion is related with the release of an anion vacancy in the neighbourhood of the I.V. dipole formed by the impurities. Because of small size-effect related with incorporation of OH^- in the anion sublattice, the formation of ion-molecules is geometrically possible in alkali chloride crystals (Stoebe). On the other hand, the divalent impurities tend to segregate preferentially near the dislocation lines, and the excess anion vacancies are responsible for the positive charge of the dislocations. The $T_M^c(c_{OH})$-dependence is similar to the concentration dependence of T_M^c characteristic for "decorated" dislocations in NaCl:Me^{++} crystals (Suszyńska et al. 1984, 1985) being in favour of the four-site model. In terms of this interpretation it is also understandable that the post-deformation RT-storage yields a composed band. The distribution either in the activation energy or in the preexponential factor, determing the relaxation time, reflects the formation of various geometrical configurations of the complexes mentioned above.

5. Summary

Although some details of the behaviour of the high-temperature ITC band in the AH : OH^- crystals are complicated and not well understood without additional suppositions, it has been shown that dislocations are important for the characteristics of this band. Following the present experiments and in accord with our previous identification, band C results from an induced polarization phenomenon related with the relative displacement of the dislocation core and the Cottrell-like atmospheres. The direction of the changes induced by any external "field" corresponds well with the dislocation charge. This effect has been confirmed with a series of preliminary experiments performed for super pure NaCl and KCl crystals irradiated with ionizing rediation. In agreement with the literature data concerning the dislocation charge (Davidge; Dupuy et al.), deformation-induced effects in heavily-γ-irradiated crystals resemble the effects reported in the present work for OH^- containing crystals. Identification of the pertinent polarization species, charge "sources" and trapping sites should be a subject of further investigations.

References

Bucci, C., Fieschi, R.: Phys. Rev. Letters **12** (1964) 16

Bucci, C., Fieschi, R., Guidi, G.: Phys. Rev. **148** (1966) 816

Capelletti, R., Gainotti, A.: J. Phys. **37**, C.7 (1976) 316

Capelletti, R., Mora, C., Zecchi, E., Carranza-Torres, O.: Rad. Effects **75** (1983) 93

Capelletti, R., Gainotti, A., Suszyńska, M.: Proc. 5-th Intern. Symp. Electret Heidelberg, IEEE New York 1985, p. 151

Cottrell, A. H., Bilby, B. A.: Proc. Roy. Soc. A **62** (1949) 49

Davidge, R. W.: Phil. Mag. 8 (1963) 1369

Dupuy, C. H. S., Schaeffer, B.: Proc. Brit. Ceram. Soc. **6** (1966) 257

Fritz, B., Lüty, F., Anger, J.: Phys. Z. **174** (1960) 240

Kessler, A.: Czech. J. Phys. B **19** (1969) 689

Müller, P.: phys. stat. sol. (a) **23** (1974) 165

Stoebe, G.: J. Phys. Chem. Solids **31** (1970) 1291
Suszyńska, M., Capelletti, R.: Intern. Symp. on Glasses and Ionic Crystals, Fincken, GDR (1978), not published
Suszyńska, M., Capelletti, R., Czapelski, M., Gainotti, A.: Sci Rep's of VIMMESS **21** (1979) 141
Suszyńska, M., Macalik, B.: Acta Phys. Polon. A **62** (1982) 363
Suszyńska, M., Capelletti, R.: Cryst. Res. Technol. **19** (1984) 1385, 1489, **20** (1985) 1363
Suszyńska, M., Capelletti, R.: Proc. of Yamada Confer. IX on Dislocations University of Tokyo Press (1985) 489
Whitworth, R. W.: Adv. Phys. **24** (1975) 203

(Received May 27, 1987)

Authors' addresses:

Dr. M. Suszyńska
Institute of Low Temperature and Structure Research, Polish Academy of Sciences
50 950 Wrocław, Poland

Dr. R. Capelletti
Istituto di Fisica dell'Universita
48 100 Parma, Italy

Book Review

P. A. KRUPCHITSKY

Grundlagenforschung mit polarisierten langsamen Neutronen

Springer-Verlag Berlin Heidelberg New York London Paris Tokyo, 1987, 138 Seiten, 42 Abbildungen, 14 Tabellen. DM 87.00. ISBN 3-540-16996-2

Die von KRUPCHITSKY verfaßte Monografie zur Grundlagenforschung mit polarisierten langsamen Neutronen kann als Weiterführung der Arbeit von ABOV u. a. (Polarisierte langsame Neutronen, 1966) angesehen werden. Da sich aber gerade in den letzten 20 Jahren die experimentelle Technik der Erzeugung, der Streuung und des Nachweises polarisierter Neutronen stürmisch entwickelt hat und sich die Originalarbeiten zu dieser Thematik auf zahlreiche Zeitschriften und viele Jahrgänge verteilen, wird durch diese Monografie eine merkliche Lücke auf dem Gebiete der Neutronenphysik geschlossen. Das Buch ist nicht nur für die auf dem Gebiet tätigen Spezialisten nützlich, sondern kann insbesondere auch den wissenschaftlichen Mitarbeitern und Studenten empfohlen werden, die auf dem Gebiete der Kernphysik oder der Elementarteilchenphysik arbeiten. Ihnen wird bei der Lektüre ein Einblick gewährt in die einzigartigen experimentellen Möglichkeiten, welche die Ausnutzung der Präzession des Neutronenspins in magnetischen Feldern für die Lösung fundamentaler Probleme der Kern- und Elementarteilchenphysik bietet.

Intensive Strahlen polarisierter langsamer Neutronen werden heute in breitem Maße für festkörperphysikalische, anwendungsorientierte Forschungen eingesetzt. Die Richtung wird in der Monografie bewußt nicht behandelt.

In der Monografie werden zunächst Fragen der Erzeugung und des Nachweises der polarisierten Neutronen sowie das Verhalten des Neutronenspins in Magnetfeldern behandelt. Dabei wird der modernste Stand der experimentellen Technik berücksichtigt. Die experimentellen Ergebnisse werden stets mit den Aussagen der Theorie verglichen. Anschließend werden die fundamentalen Probleme behandelt:

- Anisotropie und zirkulare Polarisation von γ-Strahlen, die von Atomkernen nach Einfang polarisierter Neutronen emitiert werden;
- Quantenmechanische Symmetrieeigenschaften und die Verletzung von Erhaltungssätzen bei der schwachen Wechselwirkung;
- Der Zerfall plarisierter Neutronen;
- Messung des elektrischen Dipolmoments des Neutrons;
- Anisotropie der Winkelverteilung der Produkte der Kernspaltung, die durch polarisierte Neutronen ausgelöst wird.

In der Monografie von KRUPCHITSKY werden die Probleme der Grundlagenforschung sowohl vom experimentellen als auch vom theoretischen Standpunkt in übersichtlicher Form behandelt. Die Anschaulichkeit wird durch eine gute Auswahl von Abbildungen gefördert. Die theoretische Fassung der Probleme paßt sich gut der komprimierten Form des Buches an.

H. OEHLER

Cryst. Res. Technol. | 23 | 1988 | 2 | 209–215

M. SRINIVAS, LALITHA SIRDESHMUKH

Department of Physics, Kakatiya University,
Warangal, India

Dielectric Studies on Pure and Doped $Sr(NO_3)_2$

Single crystals of pure and Na^+-doped strontium nitrate were grown by slow evaporation of aqueous solution. Systematic measurement of dielectric constant (ε) and loss ($\tan \delta$) have been carried out in the frequency range of 100 Hz to 100 kHz at temperatures ranging from room temperature to 420 °C. A.C. conductivity (σ) is obtained from the data on ε and $\tan \delta$. ε, $\tan \delta$, and σ were found to show anomalies around 300 °C. The results are discussed in the context of order-disorder phase transition.

1. Introduction

The measurement of several physical properties of univalent nitrates has indicated order-disorder phase transition associated with orientation disorder of the NO_3^- group (BADR et al.). The possibility of orientation disorder in divalent nitrates was first indicated by neutron diffraction studies of HAMILTON. Subsequently BADR and KAMEL have studied d.c. resistivity and dielectric constant of $Pb(NO_3)_2$, $Ba(NO_3)_2$ and $Sr(NO_3)_2$. They have observed anomalies in the high temperature region. This is further supported by differential thermal analysis (DTA) and thermomechanical analysis (TMA) curves obtained by them. They have attributed the anomalies to the order-disorder phase transition. BADR and KAMEL have used polycrystalline samples. Their dielectric studies pertain to the measurement of dielectric constant at a single frequency of 1 kHz. It is well established that the measurement of dielectric properties only at low frequencies gives spurious results due to the effect of space charge polarization. For quantitative estimation it is also necessary to take into account the density of the packing and correct the data on polycrystalline samples for the crystal density. In an earlier communication one of the present authors (SIRDESHMUKH) has discussed these effects on the dielectric properties of polycrystalline samples of alkaline earth nitrates and lead nitrates at room temperature. In view of the reports of possible order-disorder phase transformation, it was felt worth-while to carry out detailed systematic studies on dielectric properties of divalent nitrates. As a representative crystal of the system, $Sr(NO_3)_2$ was chosen for the present study to be followed by other members.

In the present communication we report the results of measurement of dielectric constant (ε) and loss ($\tan \delta$) on single crystals of $Sr(NO_3)_2$. Measurements have been done in the frequency range from 100 Hz to 100 kHz at temperatures ranging from room temperature to 420 °C. From the data on ε and $\tan \delta$, dielectric conductivity (σ) has been evaluated. As an additional evidence to understand the process of conduction, measurements have been made on Na^+-doped samples and the results are compared with those for pure crystal. The results are discussed in the light of existing data.

2. Experimental

Single crystals of pure strontium nitrate and with 1 mole% sodium nitrate were grown by slow evaporation of saturated aqueous solutions. The starting materials were analar grade salts obtained from Riedel De Haen (FRG). Well-formed and highly transparent crystals were obtained. Crystals were cut to dimensions of 6 × 6 × 2 mm^3 and polished. They were annealed at 200 °C for several hours. Aluminium foils were fastened on either side of the samples for good electrical contact.

The dielectric constant (ε) and loss ($\tan \delta$) were measured using a GR 1620 A capacitance measuring ensemble in conjunction with a laboratory-built three-terminal cell. The cell capacitance was found to be less than 1 pf for the entire range of measurements. For frequencies higher than 10 kHz an external oscillator (Agronic 72) was employed. The input to the heater was fed through a stabilizer and the temperature was steady within 1 K. The temperature was measured using a copper-constantan thermocouple and a digital panelmeter. The accuracy in the measurement of ε is 2% and in loss 5%. Since the crystals are hygroscopic they were kept in the dielectric cell at 100 °C for a few hours before taking measurements. Measurements have been carried out in the frequency range from 100 Hz to 100 kHz in the temperature range from room temperature to 420 °C.

3. Results and discussion

The variation of dielectric constant with temperature for various frequencies is shown in Figures 1 and 2 for pure and Na^+-doped crystals, respectively. For both the samples a slow and frequency independent increase is observed up to about 200 °C. Beyond

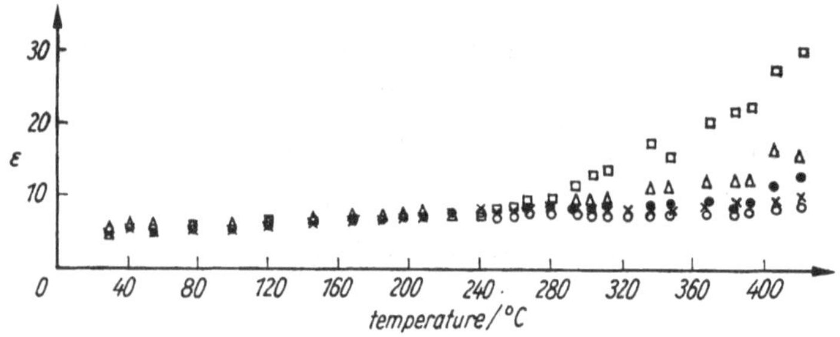

Fig. 1. Variation of ε with temperature for $Sr(NO_3)_2$. ○ 100 kHz, × 50 kHz, ● 20 kHz, △ 10 kHz, □ 5 kHz

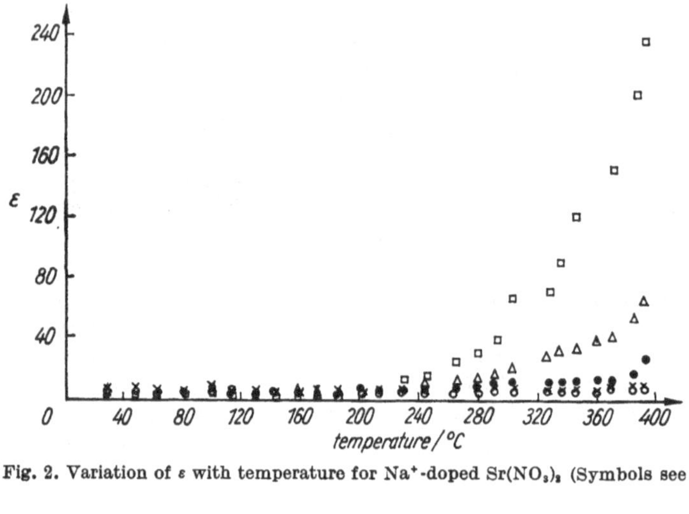

Fig. 2. Variation of ε with temperature for Na^+-doped $Sr(NO_3)_2$ (Symbols see Fig. 1)

that, a frequency dependent increase is observed. The increase in capacitance and dielectric constant at low frequencies is generally believed to be due to space charge polarization (RAO, SMAKULA). The frequency independent slow increase indicates that the space charge contribution is negligible up to 200 °C. The increase is due to electronic and ionic polarization. The effect of crystal imperfections begins to show up beyond 200 °C. The increase is faster at lower frequencies. For frequencies > 50 kHz the variation remains slow and almost frequency independent up to the maximum temperature of present measurements. It can be seen that the rate of increase with temperature at a given frequency is higher for the doped sample. The value of ε at room temperature is 5.18. It agrees with the value obtained by earlier workers (SIRDESHMUKH).

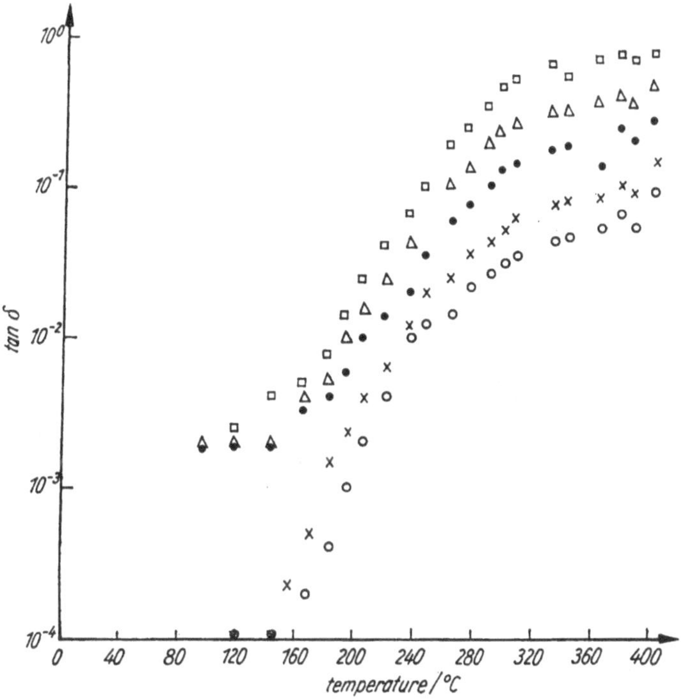

Fig. 3. Variation of tan δ with temperature for $Sr(NO_3)_2$ (Symbols see Fig. 1)

Figures 3 and 4 show the temperature variation of loss at various frequencies for pure and doped samples, respectively. For the pure crystal loss is negligible up to 100 °C. A fast increase is observed from 100 °C to 200 °C. Beyond 200 °C a slow rise accompanied by a broad peak centred around 320 °C is observed. It can be seen that the variation is similar for the doped sample with some additional unresolved peaks below 100 °C. It is to be noted that there is no shift of the broad hump around 320 °C with either temperature or frequency in both the figures.

The complex dielectric constant for a crystal is expressed as $\varepsilon^* = \varepsilon - i\varepsilon'$. The loss angle $\tan \delta = \varepsilon'/\varepsilon$. The loss ($\tan \delta$) consists of two contributions (LIDIARD) as follows

$$\tan \delta = \frac{4\pi\sigma}{\omega\varepsilon} + \frac{\varepsilon_S - \varepsilon_\infty}{\varepsilon_S + \varepsilon_\infty\omega^2\tau^2} \tag{1}$$

where ε_S and ε_∞ are the static and high frequency dielectric constant, respectively; σ is the conductivity, τ is the relaxation frequency and ω is the angular frequency. In

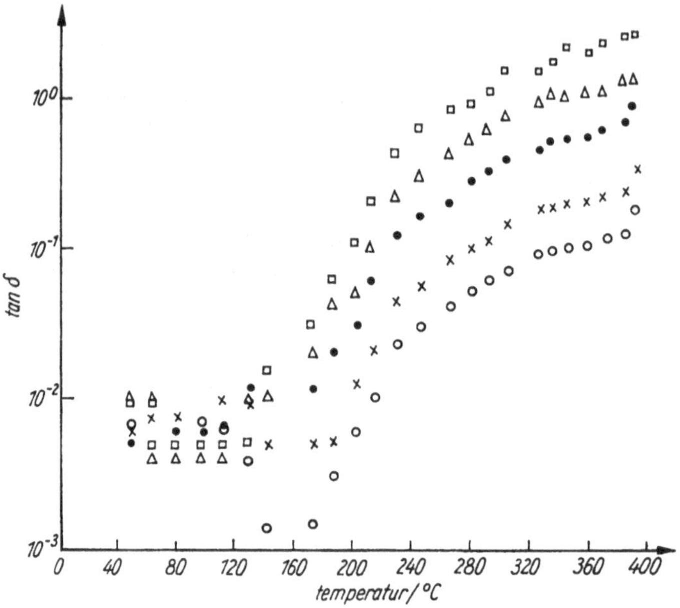

Fig. 4. Variation of tan δ with temperature for Na^+-doped $Sr(NO_3)_2$ (Symbols see Fig. 1)

equation (1) the first term represents the loss due to d.c. conduction and the second the loss due to dipolar effects. Figures 5 and 6 show log-log plots of tan δ vs frequency at different temperatures. The linear variation at temperatures indicated show that the loss is essentially due to conduction and the relaxation effects are negligible in pure crystals. For the doped samples the unresolved peaks around room temperature and up to 100 °C indicate relaxation effects. But the peaks could not be analysed as they were very weak.

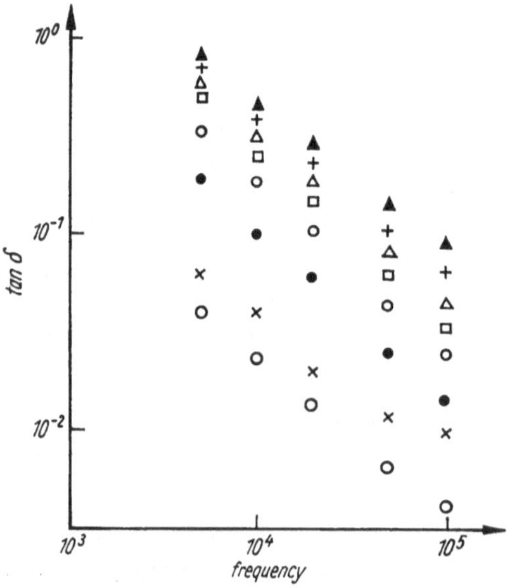

Fig. 5. Variation of tan δ with frequency for $Sr(NO_3)_2$. °C: ▲ 406, + 383, △ 346, □ 311, ○ 294, ● 267, × 240, ○ (below) 223

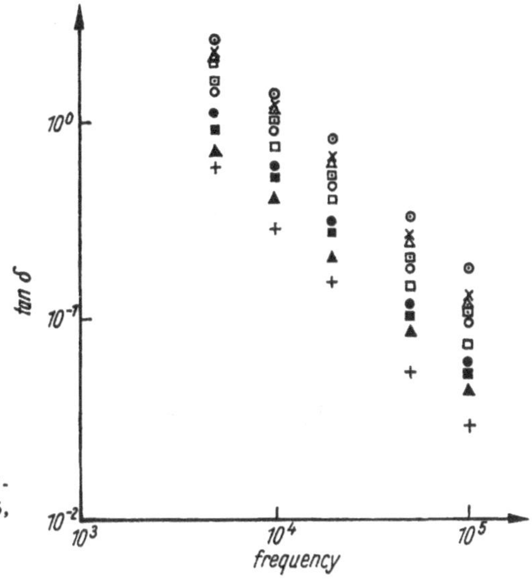

Fig. 6. Variation of tan δ with frequency for Na^+-doped $Sr(NO_3)_2$. °C: ⊙ 392, × 385, △ 370, ⊡ 345, ○ 326, □ 303, ● 291, ■ 280, ▲ 265, + 245

σ was evaluated from the relation $\sigma = \varepsilon_0 \varepsilon \omega \tan \delta$. ($\varepsilon_0$: vacuum dielectric constant). Figure 7 shows plots of log σ vs reciprocal temperature for the pure and doped samples. The curve for doped sample is shown for the highest frequency of the present measurements since frequency dependence was observed at low frequencies at higher temperatures. Up to about 100 °C where temperature has little influence, frequency dependent conductivity is observed for both the samples. The variation is more pronounced for the doped sample. Beyond 100 °C up to about 200 °C frequency independent conductivity is observed. It is clear from the diagrams that higher conductivity is indicated for the doped samples in the entire range of temperature of present measurements. The conductivity up to 420 °C is identified as extrinsic conductivity. It can be seen that the change in the slope is indicated at 200 °C and again around 320 °C. BADR and KAMEL attribute the anomalies observed at 325 °C in DTA and TMA curves to the rotation of NO_3^- group. They have related the sudden increase in dielectric constant from 200 °C onwards to the rotation of NO_3^- group leading to a solid state order-disorder transformation. The peak observed is associated with increase in thermal expansion coefficient. BADR and KAMEL believe that thermally activated reorientational processes are accompanied by loosening or weakening of ionic bonds leading to formation of dipoles. The partial rotation of the dipole may lead to further expansion of the lattice. SRINIVASAN has measured the thermal expansion coefficient of $Sr(NO_3)_2$ over the temperatures ranging from 75 °C to 525 °C. His result indicates a linear variation up to ≈ 200 °C and a higher rate of increase at higher temperatures beyond 200 °C. In the present work the change in the slope of σ vs $1/T$ curves around 200 °C and a sudden change at 320 °C could be attributed to the onset of the process involving orientational disorder at 200 °C and completion of the process at about 320 °C. The broad peaks in tan δ vs temperature curves in Figures 3 and 4 are perhaps the result of the reorientation process. It may be observed that there is no change in the position of the humps with frequency. This is characteristic of orientational disorders unlike the relaxation effects which show a shift with temperature and frequency (DANIEL).

Several reports are available in literature on the electrical conductivity of fluorite type crystals (HAYES) and those with rutile structure (PARK, NOWICK). The presence

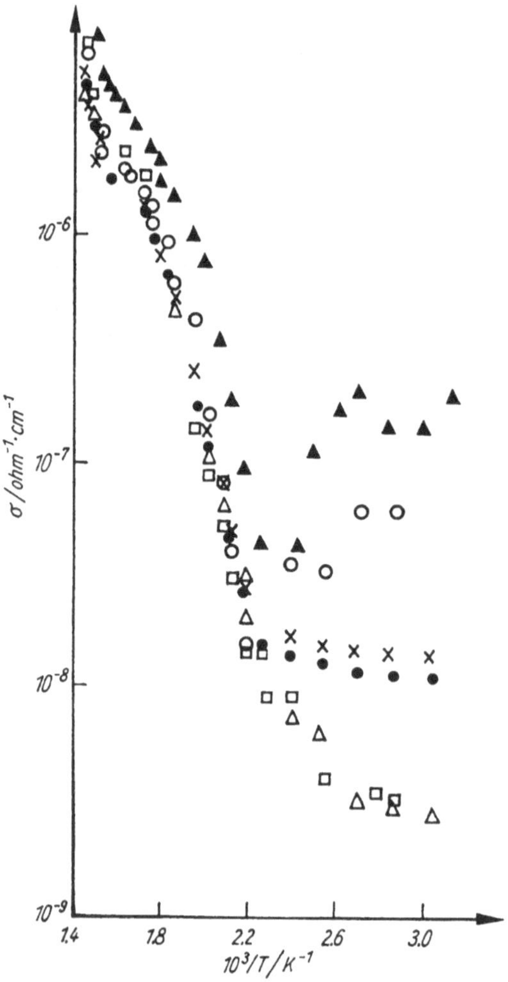

Fig. 7. Variation of σ against reciprocal temperature for $Sr(NO_3)_2$ and Na^+-doped $Sr(NO_3)_2$. kHz: ▲ 100 (with Na^+-doped) (the other symbols see Fig. 1). The units on the ordinate are to be multiplied by 10^{-2}

of negative ion vacancies and anion Frenkel defect mechanism in the case of CaF_2 and other fluorite type crystals is well-established (URE). This is shown to be the prominent defect in the chemically similar system MnF_2 with rutile structure (PARK, NOWICK). A close similarity exists in the structure of alkaline earth nitrates and fluorite structure (MICHARD). The defect mechanism in $Sr(NO_3)_2$ could be the anion-Frenkel disorder as in the case of CaF_2 and MnF_2. A strong evidence for such a conclusion is drawn from the ESR studies (TAGAYA). The existence of negative ion vacancies in $Sr(NO_3)_2$ has been established from ESR studies on X-ray irradiated $Sr(NO_2)_3$. He has observed F centres in which electrons are trapped by NO_3^- vacancies in pure crystals. In the present measurements the nominally pure sample is likely to contan -ve ion vacancies. These are enhanced by doping of aliovalent impurity. The major diefect seems to be anion-Frenkel disorder. But the process of conduction is rendered complicated by the reorientational disorder at higher temperatures. Since intrinsic conduction was not observed even up to 420 °C it was not possible to obtain quantitative estimate of defect formation energies by the present measurements.

4. Conclusions

The present investigation of dielectric properties on single crystals of $Sr(NO_3)_2$ show anomalies at higher temperatures. This is due to order–disorder phase transition as a result of reorientation of NO_3^- group. This result is in confirmity with the earlier results of DTA, TMA, and resistivity measurements.

AC conductivity data on pure crystals in conjunction with the results on $NaNO_3$-doped samples indicate that the charge carriers responsible for conduction in these crystals are the anion vacancies.

The authors wish to express their thanks to Professor D. B. Sirdeshmukh for his encouragement. One of the authors (MS) acknowledges the University Grants Commission, India, for the financial assistance.

References

BADR, Y. A., KARPOV, S. V., SHULTIN, A. A.: Sov. Phys. Solid State **15** (1974) 18
BADR, Y. A., KAMEL, R.: J. Phys. Chem. Solids **41** (1980) 1127
DANIEL, V. V.: Dielectric Relaxation, London, New York 1967
HAMILTON, W. C.: Acta Cryst. **10** (1957) 103
HAYES, W.: Crystals with Fluorite structure, Oxford 1974
LIDIARD, A. B.: Encyclopedia of Physics (Edited by S. FLÜGGE) **20** (1957) 246
MICHARD, F.: D. Sc. Thesis, Univ. Paris 1973
PARK, D. S., NOWICK, A. S.: J. Phys. Chem. Solids **37** (1975) 607
RAO, K. V., SMAKULA, A.: J. Appl. Physics **36** (1965) 2031
SIRDESHMUKH, L.: Curr. Sci. **44** (1975) 960
SRINIVASAN, R.: Proc. Ind. Acad. Sci. A **41** (1955) 49
TAGAYA, Kimihito.: Jpn. J. Appl. Phys. **23** (1984) 168
URE, R. W.: J. Chem. Phys. **26** (1957) 1363

(Received June 26, 1987)

Authors' address:
Dr. Lalitha SIRDESHMUKH, M. Srinivas
Department of Physics, Katakiya University
Warangal — 506 009, India

Book Review

E. Kaneko

Liquid Crystal TV Displays: Principles and Applications of Liquid Crystal Displays

Tokyo, KTK Scientific Publishers; Dordrecht, Boston, Lancaster, Tokyo, D. Reidel Publishing Company, 1987, 307 pages, 169 figures, 29 tables, Dfl. 240.00/US $ 99.00. ISBN 90-277-2362-1

In the last 20 years the liquid crystals have developed from a scientific curiosity to an important topic with relevant worldwide practical application, the estimated output of liquid crystal displays (LCD's) in 1986 was 800 Mio units. In this situation the availability of a comprehensive survey on the many different electrooptic effects of LCD's and their applications is of great value. E. Kaneko, chief researcher of the HITACHI research laboratory has undertaken the troublesome task of reporting progress on the said subject. The author himself is actively working in this field, he has published papers and a book devoted to LCD's and therefore he is a distinguished expert in presenting a state of the art report.

The book is divided into the following chapters: 1. Introduction 2. Basic concept of LC cells 3. Color LC cells 4. Direct addressing LCD and LC TV displays 5. Some special methods for direct addressed matrix LCD's 6. Double threshold matrix addressed LCD's 7. Transistor switch addressed LC TV displays 8. Large screen projection type LCD's.

The technically interested reader can find an excellent overview on the different types of LCD's and their basic concepts. Of course, the newest types of supertwisted nematic displays (STN, OMI) are lacking since they have been developed after finishing the manuscript, however, the book is on the newest possible stand. The high standard of the physical and technical part of the theme could not be maintained in the chemically oriented sections. Unfortunately, the text is suffering from misprints and linguistic errors. The index is too small for a profitable use. Printing and establishment of the book are excellent.

Kaneko's work in the best manner is filling a gap in the present literature and will be a standard reference source for all people interested in LCD's.

D. Demus

Cryst. Res. Technol.	23	1988	2	217–222

K. J. PRATAP, M. NAGABHUSHANAM, F. M. AMANULLAH, V. HARI BABU

Department of Physics, University College of Science,
Osmania University, Hyderabad, India

Thermo emf Studies on CdTe:Sb Thin Films

Thermal transport properties have been studied on as-grown and annealed CdTe:Sb thin films of different thicknesses and concentrations of Sb. These films were prepared by sequential vacuum deposition of CdTe and Sb on clean and hot glass substrates held at 210 °C temperature at a constant rate of deposition. From these studies all the films were found to be p-type. It is also observed that the thermoelectric power increases with Sb concentration, thickness of the film and with annealing process.

Fermi-energy variation with temperature have been calculated on the basis of thermo emf data. All these results have been explained on the basis of defects introduced by Sb doping.

Introduction

Studies on doped CdTe films are interesting due to their device applications in many fields like IR detectors, photoconductors, thin film transistors (PIZZARELLO; VASS, ANDERSON; EDDALLS). Though there are reports on the transport properties (in particular with thermoelectric power) of CdTe single crystals, very little is in literature on doped CdTe thin films. Recently, MÜLLER et al. have studied the thermoelectric power (TEP) of plastically deformed p- and n-type CdTe and concluded that highly conductive regions are connected with dislocations in p–CdTe. DAS and BHATT have observed anamalous temperature dependence of thermoelectric power of p-PbTe thin films. The anomaly being invariant of TEP with temperature at low temperature and sudden decrease at higher temperature. It is the aim of this work to present experimental results on thermoelectric power of CdTe films doped with different concentrations of antimony and for different thicknesses. The effect of annealing on the above said two sets of films was also studied. These results are explained on the basis of defects introduced by Sb impurities.

Experimental

CdTe:Sb thin films were prepared by sequential vacuum deposition of CdTe and Sb on clean and hot glass substrates held at 210 °C temperature and at a constant rate of deposition in a vacuum of 5×10^{-5} torr. Annealing was done at 200 °C in vacuum for 30 mts. using a resistance heater placed just above the substrate holder. The thickness of the films was measured with multiple beam interferometer. Films of different thicknesses were prepared by controlling the evaporation time. The dopant Sb was controlled by taking different fixed amounts of Sb at atomic proportions. Thick Al electrodes were predeposited on the glass substrates served as metallic contacts with film and the TEP was measured in vacuum as a function of temperature. The TEP measurement set up consists of a massive copper heat sink and to this copper sink one end of the experimental film was clamped to maintain its temperature constant. This end (A) was the cold end of the film. To the other end of the film a mini copper heater was clamped which would be heated resistively. This end (B) of the film was heated uniformly by controlling the power supply. The temperature of the hot end of the film was measured using copper-constantan thermocouple. Maintaining the lower junction at room temperature, the temperature of the upper junction was raised slowly. The thermal emf developed across the CdTe:Sb film was measured as a function of hot end temperature. The emf values were measured with a Philips DC microvoltmeter. From the measured temperature dependence of the thermo emf, the relative thermoelectric power ΔS and absolute thermoelectric power S_F of the films were obtained.

Results

Plots of thermo emf versus temperature difference for as-grown and annealed CdTe:Sb (6.6%) films of different thickness (1020, 1840, and 2410 Å) are shown in Figure 1. It is seen from this figure that thermo emf increases non-linearly with increasing temperature difference. Further, both in as-grown and annealed films non-linearity decreases at thickness increases. It is also seen from the same figure that the emf developed for the same temperature difference is higher (i) for higher thickness films and (ii) in an annealed film. A similar variation in CdTe films doped with different concentrations of Sb was also observed (not shown).

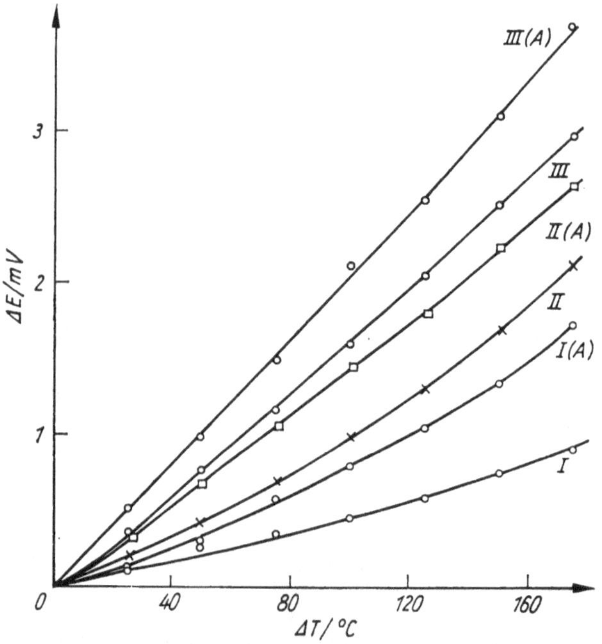

Fig. 1. Thermo emf vs temperature difference of as-grown and annealed CdTe:Sb films of different thickness

The thermo emf S_{AB} of CdTe film end (B) with respect to end (A) was found to be negative indicating that the sample is p-type ($S_{AB} = S_A - S_B$ is < 0; but $S_A > 0$ and hence S_B is positive, hence, p-type).

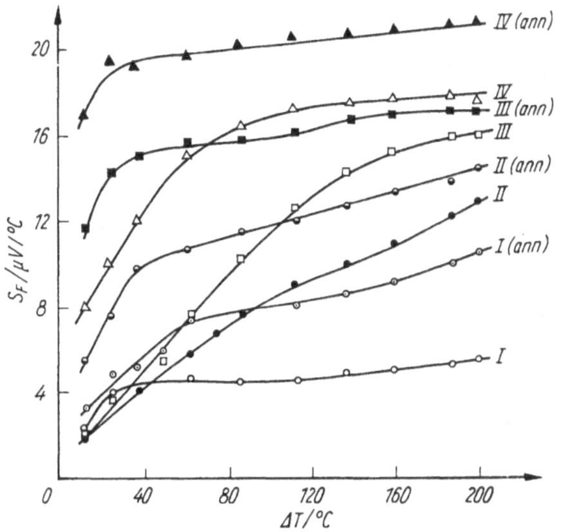

Fig. 2. Thermo electric power vs temperature of CdTe:Sb films of different thickness

Figure 2 shows the graph of thermoelectric power, S_F against temperature for as-grown and annealed CdTe:Sb (6.6%) films of different thicknesses. It is seen from this figure that S_F increases (i) with increasing thickness of the film and (ii) with annealing process. It is also seen that the rise in S_F is non-linear in as-grown and almost linear in annealed films.

Figure 3 shows the plots of thermoelectric power S_F against temperature for as-grown and annealed CdTe:Sb films with different concentrations of Sb (4.3, 6.6, 8, and 10%). From this figure it is seen that S_F increases (i) with increase in Sb concentration and (ii) with annealing. From the figure it is also observed that all the curves have very slow linear rise in S_F with temperature.

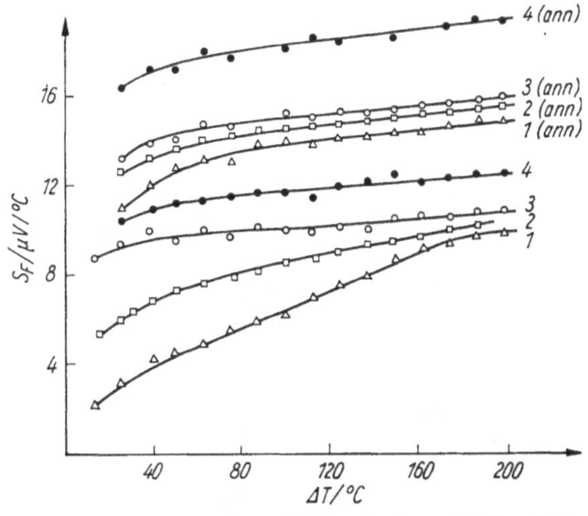

Fig. 3. TEP vs temperature of CdTe:Sb films of different Sb concentrations

Discussion

For a non-degenerate p-type semiconductor the thermoelectric power S_F is given by the expression (4)

$$S_F = \frac{k}{e}\left(\frac{5}{2} + \alpha + \frac{E_F - E_v}{kT}\right) \tag{1}$$

$$= \frac{k}{e}\left(\frac{5}{2} + \alpha + \ln\frac{Nv}{p}\right), \tag{2}$$

where

$$Nv = 2\left(\frac{2\pi m_p^* kT}{h^2}\right)^{3/2}$$

E_F is the Fermi energy, E_V is the energy of the top of the valence band, k is the Boltzmann constant, e is the charge of the electron or hole, T is the absolute temperature, p is the hole concentration and α is a constant defined by Müller et al.

$$\tau(E) = \tau_0 E^\alpha \tag{3}$$

where τ the relaxation time, is considered to be a function of energy E, and τ_0 is a constant. N_v is the effective density of states in the valence band. If scattering by acoustic or non-polar optical phonons is predominant α should be equal to -0.5 (HERRING; DEVLIN). Since $N_v > p$, the expression inside the bracket is positive and hence the sign of the thermoelectric power determined the sign of the carriers. It is found in all CdTe:Sb films under investigation S_F is positive throughout the temperature range suggesting the films to be p-type in nature.

It is well known (CHOPRA) that the defect concentration in thin films (point defects, point clusters, dislocations and grain boundaries) is much larger than in the bulk. This concentration is a function of thickness of the film, decreasing with increasing thickness upto a certain critical thickness (around 3000 Å, depending upon the material) and then reaching saturation for higher thickness. Since the CdTe:Sb films under study have thicknesses below 3000 Å, one can expect the variation of defects with thickness. These defects scatter the charge carriers and can lower the thermoelectric power. The decrease in thermoelectric power with decrease in thickness of the film could therefore be due to structural defects. With annealing S_F increases and this probably is due to decrease in the concentration of defects with annealing.

In all annealed films the TEP is positive and increases linearly with low rate. It is not strongly temperature dependent in the temperature range of present investigation.

The values of $E_F - E_v$ for all the films studied are calculated using the experimental data of S_F and $\alpha = -0.5$ in the equation (1). The variations of $E_F - E_v$ with temperature for all the as-grown and annealed films are shown in Figure 4 and the corresponding slopes of the curves are given in Table 1. It is clearly seen from the table that $\delta E_f/\delta T$ values are more for as-grown films when compared with their respective annealed films. This indicates a significant reduction of the carrier density with increasing temperature; whereas in annealed films reduction is not much. In addition, different workers (LEGRES et al.; AGRINSKAYA et al.) have shown that acceptor clusters are formed invariably in annealed compound semiconductors. Added to this the Sb impurities may also create acceptor defects in CdTe lattice. These two effects probably, reduce the electric power with temperature. The slightest rise in S_F with T may be due to the variation of α with temperature and thickness of the film. In annealed films the visible decrease in the slope of the plots (S_F vs T) with increase in film thickness

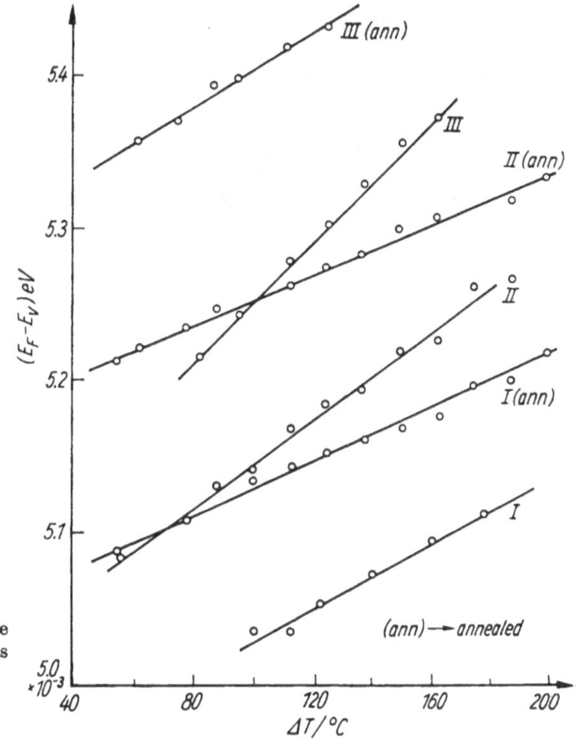

Fig. 4. Variation of $(E_F - E_V)$ with temperature of CdTe:Sb (annealed and un-annealed) films of different thickness

Table 1

Thickness Å	$\frac{\delta E_f}{\delta T}$ μv/k⁰		γ (ev k⁻¹)
	as-grown	annealed	
1020	0.091	0.0073	-2.05×10^{-4}
1840	0.010	0.0088	-1.85×10^{-4}
2410	0.019	0.0077	-1.75×10^{-4}

requires $(\alpha = \gamma/k)$ to decrease with increase in thickness. A similar type of variation of γ was observed by NAIDU and REDDY. Values corresponding to the annealed films are also given in Table 1.

In CdTe films doped with 4.3%, 6.6%, 8%, and 10% Sb, the variation on S_F with temperature (Fig. 3) is similar both in as-grown and annealed films except for the film doped with 4.3% of Sb. This may be due to precipitation or cluster formation at high concentrations of doping.

It has also been observed that the variation of S_F is independent on temperature in both as-grown and annealed films with concentrations more than 6%.

The authors are thankful to the Head, Department of Physics, Osmania University, Hyderabad-7, for providing necessary facilities and constant encouragement. One of the authors (Pratap) is thankful to UGC, New Delhi, for awarding FIP. Also one of the authors (FMA) is thankful to the Principal and Management, Deccan College of Engineering and Technology, Hyderabad, for encouraging to carry out the research work.

References

AGRINSKAYA, N. V., ARKADEVA, E., MATVEER, O. A.: Sov. Phys. Semicond **4** (1970) 347
CHOPRA, K. L.: thin film phenomena, New York 1969, p. 246
DAMODARA DAS, V., BHAT, Seetharama: J. Appl. Physics **54** (11) (1983) 6641–6645
DEVLIN, S. S.: in: Physics and Chemistry of II–VI compounds, Ed. M. AVEN and J. S. PRENER, Amsterdam 1967, p. 448
EDDALLS, D. V.: Infrared Phys. **16** (1976) 47
HERRING, C.: Phys. Rev. **96** (1954) 1163
LEGRES, R., MARFAING, Y., TRIBOULET, R.: J. Phys. Chem. Solids **39** (1978) 179
MÜLLER, H., HAASEN, P., SCHRÖTER, W.: phys. stat. solidi. (a), **80** (1983) 223
PIZZARELLO, F. A.: J. Appl. Physics **38** (1967) 1752
SRINIVASULU NAIDU, B., JAYARAMA REDDY, P.: Thin Solid Films **61** (1979) 379
VASS, R. W., ANDERSON, R. M.: J. Appl. Physics **45** (1974) 3463

(Received May 29, 1987)

Authors' address:

Prof. V. HARI BABU, K. J. PRATAP, M. NAGABHUSHANAM, F. M. AMANULLAH
Department of Physics, University College of Science,
Osmania University
Hyderabad — 500007, India

Cryst. Res. Technol.	23	1988	2	223–229

L. Malicskó, Á. Barna, R. Scholz

Research Laboratory for Crystal Physics of the Hungarian Academy of Sciences, Budapest, and
Research Institute for Technical Physics of the Hungarian Academy of Sciences,
Budapest Hungaria, and
Institute for Solid State Physics and Electron Microscopy
of the Academy of Sciences of GDR,
Halle, GDR

Investigations on the High Speed Ion Beam Thinning of Lithium Niobate Single Crystals

1. Introduction

Because of some unfavourable behaviour of the lithium niobate (LN) single crystals widely used in the microelectronics and laser techniques, it is not easy to prepare foils of them to be investigated by transmission electron microscopy (TEM). For the final thinning of electron transparent foils of similar materials (e.g. minerals) ion beam etching has sucessfully been used for about 30 years. But this method may also imply several disturbing effects, such as changes in the composition and in the crystallinity of the surface layer (Goodhew; Baretzky, Taglauer), as well as the generation of different topological artefacts (Tagg et al.).

The aim of the present investigations was to study the effects of high-speed and low-angle ion beam thinning by reflection high energy electron diffraction (RHEED) and TEM in detail on LN single crystals using a newly developed instrument (Barna).

2. Experimental

From undoped, Ti- and (Ti + Mg)-doped LN single crystals, grown by the Czochralski technique (Schmidt, Voszka; Polgár), (001) oriented discs were cut, lapped and polished with diamond paste of 1 μm grain size. The starting surfaces of the discs under investigation were then produced by further mechanical polishing with alumina pastes of 0.3 and finally 0.05 μm grain size. In this way 200 μm thick discs of 5 mm diameter were prepared from undoped crystals for the RHEED investigations, and 50 μm thin discs of 3 and 2.3 mm diameter from undoped and doped crystals for the TEM investigations.

The thick discs were then bombarded from one side with 3 to 10 keV Ar^+ ions at 100 to 200 μA ion currents (at 10 kV) in high vacuum between 7×10^{-3} and 7×10^{-2} Pa, at low incident angles of 7° down to 1.5° related to the surface tangent, for 10 to 60 minutes using a "teletwin" ion source with a beam of about 1° divergence limit and of 0.4 mm diameter (Barna). During ion bombardment the samples were rotated. The thinning rates amounted to values between 10 and 30 μm/hour at 10 kV.

Starting from the mechanically well polished state of the samples after each subsequent step their surfaces were studied by RHEED in a JEM 100 U (TEM) in order to follow their different states and changes.

Check experiments were also carried out to get to know the effect of the expected amorphization of the bombarded surface layers in the RHEED patterns. Therefore, a LN single crystal target was bombarded with 10 keV Ar^+ at 5×10^{-2} Pa and 150 μA ion current. The material sputtered from the target crystal was condensed as amorphous surface layers of thicknesses of (5 ± 1) and (2.4 ± 0.5) nm onto the well polished surface of another LN single crystal at a rate of 0.1 nm/s, as measured by a quartz monitor. The samples bombarded or covered were also annealed at 673 K in air or in high vacuum, respectively.

The 50 μm discs were finally thinned from both their sides with 7 to 10 keV Ar^+ or Kr^+ ions at low incident angles between 3 and 5° in high vacuum until they perforated. The titanium foil aperture of the sample holder was of 2 mm diameter and of 0.05 mm thickness. The samples were then investigated in a JEM 100 U equipped with a wave- length dispersive X-ray spectrometer (WDS), in a JEM 1000 D and a JEM 100 C conventional, high-voltage and high-resolution TEM, respectively.

3. Results and discussion

The results obtained by RHEED on thick LN specimens are demonstrated in Figure 1. The mechanically well polished surfaces provide the best single-crystal RHEED pat-

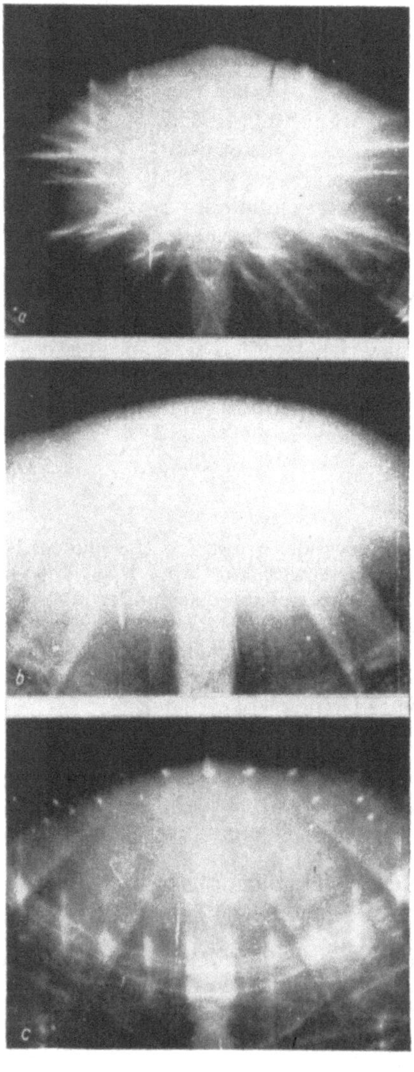

Fig. 1. RHEED patterns taken at 2° incident angle and in the same pole position from: a – a mechanically well polished surface of an undoped lithium niobate single crystal, b – the same crystal surface as in Figure 1a, after a final bombarding with 3 keV Ar^+ ions, and c – the ion bombarded surface after the thermally induced recrystallization

terns which have been obtained on LN samples (Fig. 1a). This good surface quality has the advantage that even at large incident angles of the electron beam up to 10° related to the surface tangent relatively good reflections and Kikuchi-lines appear.

After the 10 min. bombardment with 10 keV Ar^+ ions under the above mentioned conditions at small angles, however, a certain damage of the surface can already be recognized in the RHEED patterns in comparison to those obtained from the mechanically well polished surfaces. In out-of-pole positions of the samples diffusive rings in the RHEED patterns indicate that the uppermost surface layer is strikingly amorphous. In consequence of this amorphous layer in pole positions Kikuchi-lines of more or less weak contrasts only appear pointing to the single crystal state of the deeper regions of crystals. A similar decreas in contrast caused by amorphous covering layers in the channeling patterns was recently observed (Ishiguro et al.). This damage, i.e. the amorphization of the uppermost atomic layers of the surface, cannot entirely be removed by subsequent further bombardments at smaller angles and/or smaller energies down to 3 keV (Fig. 1b).

It was established that after any ion bombardment of a LN crystal surface the quality of the latter was not as good as that reached by careful mechanical polishing. If heavier Kr^+ ions were used instead of Ar^+ the surface damage under the same conditions seemed to be smaller, which is in qualitative agreement with earlier results (Cullis et al.).

According to the check experiments, the 2.4 nm thick amorphous covering layer condensed from the own material on the mechanically well polished specimen surface, results in RHEED patterns similar to those obtained after the bombardments mentioned in connection with Figure 1b. The 5 nm amorphous covering layer, however, almost entirely prevented the Kikuchi-lines emerging from the support. Thus, the thickness of the amorphous layers on the bombarded LN specimen surface can be estimated to be about 2 nm. This value is in good agreement with the thickness values recently measured for damaged layers (Howitt; Bulle-Lieuwma, Zahn; Ishiguro et al.), and which can be calculated from the empirical formula recently found (Ishiguro et al.).

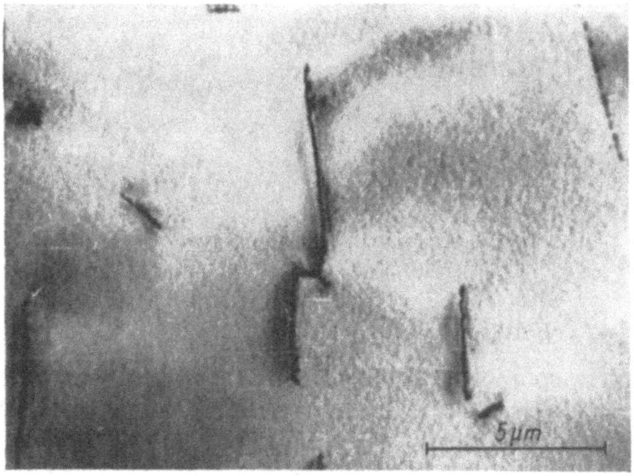

Fig. 2. Diffraction contrast image of dislocations in the surroundings of a foreign precipitated particle (not seen in this picture) taken on an ion-thinned (001) foil of a doped lithium niobate single crystal exhibiting no traces of any ion beam damages

The amorphous layers formed after ion bombardment or by extra condensations on the surface of the specimens can entirely be recrystallized by annealing for 1 hour in air or in vacuum at 673 K (Fig. 1c). In the RHEED patterns of the recrystallized surface regions extra spots also appear in addition to the matrix reflections. These

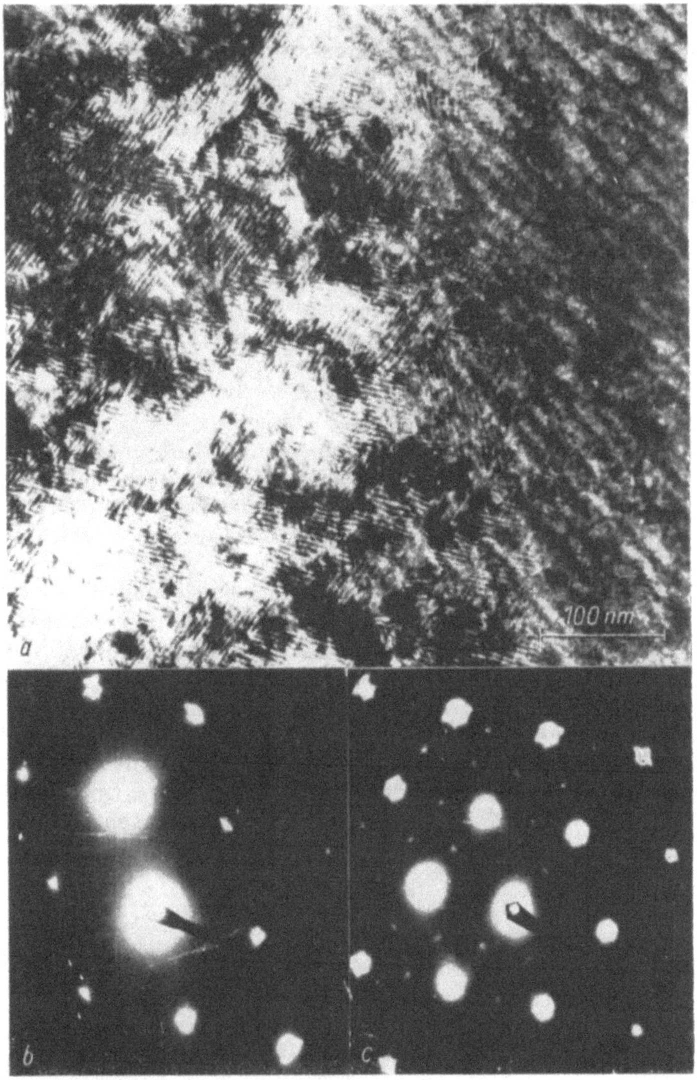

Fig. 3. Bright field image of the thin part of a Ti-doped lithium niobate single crystal foil thinned by ion beam, strikingly showing moirè-patterns in the surroundings of a bent contour (a). The corresponding SAD patterns: b – in the thicker part of the sample (for comparison), and c – in the thin part, showing satellites and other extra spots

extra spots are due to the LN crystalline surface regions implantation-modified and partially decomposed (HOFFMAN; MÜLLER, and PFLUG).

After the final ion beam thinning, a yellow or gray colouration of the TEM samples can visually be observed as a consequence of the possible compositional change, e.g. oxygen loss (HOFFMAN; RÄUBER; POLGÁR). But on carefully thinned samples no traces of any ion beam damage were detected in the TEM imaging at magnifications up to about 10000-times, not considering, of course, the possibly existing well-known flat hilly topography of the surfaces (TAGG et al.). The samples are well suitable for the study of different crystal defects. This is demonstrated in Figure 2. by the diffraction contrast image of dislocations surrounding a precipitated particle in a doped (001) LN single crystal foil.

At higher magnifications, however, hints of ion beam damages are already noticeable mainly if the ion energies and the incident angles exceed the values of 7 keV or 5 degrees, resp.: in the brigth- and also in the dark-field images moiré-patterns can be recognized which appear most strikingly along bent contours (Fig. 3a). In the associated selected area diffraction patterns satellites and other extra spots appear. These extra spots become more distinct in approaching to the thinner parts of the samples near the perforation edges (Figs 3b to c). These facts point to the presence of crystalline surface layers covering the matrix foils. It should be mentioned that the extra spots detected in the RHEED patterns of the recrystallized samples and in the SAD patterns of the respective TEM samples were found to be indentical. That means, the surface processes during ion thinning, which were followed by RHEED on thick samples, are similar to those which may take place during the final thinning of the thin TEM samples.

Figure 3a shows an interesting, quasiperiodic oriented stripe structure, too. This stripe structure can only be observed on strongly doped LN samples. Similar features were also found on LN samples finally thinned by chemical etching. Owing to the considerable contrasts these features are believed to reflect fine compositional variations, e.g. as-grown spinoidal decompositions in the crystals (CHAMPNESS, LORIMER). Such features had not been observed in lithium niobate before.

Samples thinned by ion beams up to their perforation were also investigated in a high-resolution TEM. Though covered with thin surface films they do not generally prohibit lattice fringe images to be taken of the matrix substance. Figure 4 shows lattice fringes of (110) planes of a LN crystal thinned by ion beam. Relative to figures taken of samples thinned earlier by a combined mechanical-chemical method (MALICSKÓ, SCHOLZ), Figure 4 reveals no remarkable difference, i.e. no remarkable disturbance in its quality. Of course, due to the moiré pattern of the ion-thinned sample Figure 4 displays a fluctuation in the background intensity.

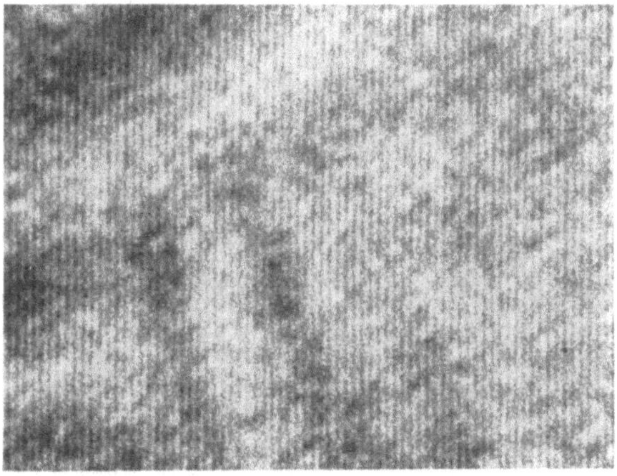

Fig. 4. (110) lattice plane fringes taken on an ion-thinned (001) LN single crystal foil (spacing: 0.258 nm)

As to the possible co-implantation of foreign substances from the surroundings into the surface layer of the samples during ion beam thinning, further check experiments were carried out by using WDS in the TEM. In our case the most important foreing substances to be expected may be: the Al as the cool cathode material and the Ti as the material of the rotating sample holder. In spite of the high energy bombardement of our specimens no traces of Al were detected by WDS investigations. Consequently, if any Al traces are present, they should be smaller than 0.1 monolayer.

In addition a thin LN sample of 3 mm in diameter was masked with a 0.05 mm thick tantalum aperture of 0.7 mm in diameter, and then thinned with 10 keV Ar^+ ions at an incident angle of 10°. In this sample a Ta-signal/square root background ratio of 15.3 was measured contrary to a similar ratio of 1.82 measured in the non-masked reference samples. From the Ta-signals a Ta coverage of 0.25 nm can be estimated for either side. This result verified our assumption that besides the Ar or Kr beam material and the residual gases (SILVA et al.), the co-implantation of other substances may also occur, which are present in the surroundings of the samples during ion milling and which can be sputtered because of the ion-bombarded area relations at given incident angles. This information must be taken into account for all special ion beam thinning procedures using masking or embedding for any purpose.

Summarizing the results the following conclusions may be drawn. In the case of LN the careful fine mechanical polishing produces a single-crystalline surface of high quality. After ion bombardment, though using very low angles and low energies, a more or less thin amorphous surface layer should be expected. This layer can recrystallize by annealing. As experimentally established, if no extra cooling is used during the TEM preparations just before the thinning is finished, in the very thin foil the temperature can rapidly increase up to a few hundred centigrades possibly initiating a recrystallization. If so, the process results in a crystalline cover layer causing moiré-patterns in TEM images. It should be noted that samples with pronounced moiré-patterns did not show any traces of excess dislocation formation. But due to the implantation of the bombarding ions and other possible co-sputtered and recondensed species (e.g. mask substances) composition changes of the cover layers of samples must also be taken into account. Because ion thinning damage effects may radically be reduced by well-chosen experimental conditions, this method can be successfully applied to the preparation of TEM samples from lithium niobate and similar crystals which are hard to handle otherwise.

The authors like to thank to Prof. Dr. J. Heydenreich for his valuable discussions. This work was supported by the State Office for Technical Development (OMFB) and the Academy Research Foundation (AKA) in Hungary.

References

BARETZKY, B., TAGLAUER, E.: Surface Sci. **162** (1985) 996

BARNA, Á.: in: Proc. 11th European Congr. on Electron Microscopy, Budapest 1984, p. 107

BULLE-LIEUWMA, C. W. T., ZAHN, P. C.: in: Proc. XIth Int. Congr. on Electron Microscopy, Kyoto 1986, p. 351

CHAMPNESS, P. E., LORIMER, G. W.: in: Electron Microscopy in Mineralogy. Ed. H.-R. WENK, Berlin 1976, pp. 174 to 270

CULLIS, A. G., CHEW, N. G., HUTCHINSON, J. L.: Ultramicroscopy **17** (1985) 203

GOODHEW, P. J.: Specimen Preparation in Materials Science, Amsterdam 1973, pp. 83 to 91

HOFFMAN, S.: in: Secondary Ion Spectrometry. Eds. A. BENNINGHOVEN, J. GIBER, J. LÁSZLÓ, M. RIEDEL and H. M. WERNER, Berlin 1982, pp. 186 to 200

HOWITT, D. G.: J. Electron Microscopy Technique **1** (1984) 405

ISHIGURO, T., SUZUKI, T., SUZUKI, N., OZAWA, M., in Proc XIth Int. Congr. on Electron Microscopy, Kyoto 1986, p. 353

MALICSKÓ, L., SCHOLZ, R.: phys. stat. sol. (a) **78** (1983) K1

MÜLLER, E., PFLUG, G.: in: Veröffentlichungen zur 11. Tagung Elektronenmikroskopie, Dresden 1984, Vol. 1. p. 312

Polgár, K.: Thesis, Budapest 1980
Räuber, A.: in: Current Topics in Materials Science. Vol. I. Ed. E. Kaldis, Amsterdam 1978, pp. 484 to 601
Schmidt, F., Voszka, R.: Cryst. Res. Technol. **16** (1981) K127
Silva, E., Robinson, J. W., Northwood, D. O.: J. Materials Sci. Lett. **5** (1986) 343
Tagg, M. A., Smith, R., Walls, J. M.: J. Materials Sci. **21** (1986) 123

(Received June 13, 1987)

Authors' addresses:

Dr. L. Malicskó
Research Laboratory for Crystal Physics, Hungarian Academy of Sciences
POB 132
H-1502 Budapest, 112, Hungary

Dr. Á. Barna
Research Institute for Technical Physics, Hungarian Academy of Sciences
POB 76
H-1325 Budapest, Újpest I, Hungary

Dr. R. Scholz
Institut für Festkörperphysik und Elektronenmikroskopie der
Akademie der Wissenschaften der DDR
POB 250
Halle
DDR-4050

Book Review

V. F. Gantmakher, Y. B. Levinson

Carrier Scattering in Metals and Semiconductors

Modern Problems in Condensed Matter Sciences Vol. 19

North-Holland: Amsterdam, Oxford, New York, Tokio 1987, 459 Seiten. Dfl. 280.00. ISBN 0-444-87025-3

Transportprozesse in Metallen und Halbleitern sind ein umfangreiches Gebiet physikalischer Forschung und von großer technischer Bedeutung. Eine theoretische Behandlung und Berechnung von Transportgrößen ausgehend von ersten Prinzipien stößt jedoch auch heute noch auf große Schwierigkeiten. Für nahezu ideale Kristalle, deren gitterperiodische Struktur nur geringfügig gestört ist, erlaubt das Quasiteilchenkonzept die einheitliche Beschreibung einer Vielzahl von Transportgrößen. Im Rahmen dieses Konzepts wird die Relaxation eines Test-Quasiteilchens untersucht, das an Gitterdefekten oder Phononen gestreut wird. Die Bandstruktur des Idealkristalls bestimmt dabei die Energie-Impuls-Beziehung des Quasiteilchens. Aus den Relaxationszeiten des Quasiteilchens lassen sich die interessierenden Transportgrößen ermitteln. Unberücksichtigt bleiben bei dieser Vorgehensweise alle quantenmechanischen Interferenzeffekte, die für das Phänomen der Lokalisierung wesentlich sind. Auch kann die genannte Methode die numerisch äußerst aufwendige Lösung der Boltzmanngleichung nicht ersetzen. Zweifellos liegt der Vorteil des Quasiteilchenkonzepts jedoch darin, daß mit relativ einfachen Mitteln eine physikalisch durchsichtige Beschreibung der Transportprozesse in Metallen und Halbleitern gegeben werden kann.

Die vorliegende Monographie ist die erste in sich geschlossene Darstellung der Quasiteilchenstreuprozesse in Metallen und Halbleitern. In Kap. 1 und 2 wird ein knapper Überblick über die Bandstruktur von Metallen und Halbleitern gegeben, Grundlagen der Streutheorie zusammengestellt und das Quasiteilchenkonzept erläutert. In den folgenden Kapiteln werden in systematischer Folge die verschiedenen Streumechanismen, die das Transportverhalten in Metallen und Halbleitern bestimmen, untersucht. Streuprozessen an Phononen sind die Kapitel 3, 4, 5 und 8 gewidmet, Elektron-Elektron-Wechselwirkung behandelt Kap. 6 und die Streuung an Punktdefekten, Versetzungen und Oberflächen wird in den Kap. 9–11 behandelt. Gesonderte Kapitel sind den Spin-Umklapp-Streuprozessen und der Streuung bei Vorhandensein eines Magnetfelds gewidmet (Kap. 13–15).

Das Buch wendet sich an Theoretiker wie auch Experimentatoren, die auf dem Gebiet der Festkörperphysik arbeiten. Es kann aber auch Physik-Studenten der höheren Studienjahre mit der Spezialisierungsrichtung Festkörperphysik empfohlen werden.

W. John

Cryst. Res. Technol. | 23 | 1988 | 2 | 231–236

N. ARUTYUNOV, A. BALTENKOV, I. SEGAL

Ministry of Geology USSR, Middle-Asiatic Institute
of Geology and Mineral Raw Materials, Tashkent, USSR

Anisotropy Effect of High-Momentum of ADAP in Diamond-like Semiconductors¹)

A high-momentum component of angular distributions of annihilation photons (HMC ADAP) has been studied experimentally and theoretically in single crystals of germanium, silicon and gallium arsenide and in a single crystal and polycrystal silicon. The model oe the process of positron annihilation with electrons of ion cores has been suggested and the calculations of HMC ADAP curves have been performed for crystallographic directions [111], [110] and [100] of a dimond-like lattice. It has been established that ADAP anisotropy characteristics in a single crystal of germanium are slightly dependent on the temperature in the range of $\Delta T = 300 \div 930$ K. It has been concluded that HMC ADAP anisotropy is caused by annihilation of thermalizod positrons with electrons of outer shells of ion cores in crystal lattices of dimond-like semiconductors.

Экспериментально и теоретически исследован широкий компонент угловых распределений аннигиляционных фотонов (ШК УРАФ) в монокристаллах германия, кремния и арсенида галлия, а также в моно- и поликристаллическом кремнии. Предложена модель процесса аннигиляции позитронов с электронами ионных остовов и проведены расчеты кривых ШК УРАФ для кристаллографических направлений [111], [110] и [100] решетки типа алмаза. Установлено, что характеристики анизотропии УРАФ в монокристаллическом герамнии относительно слабо зависят от температуры в интервале $\Delta T = 300 \div 930$ К. Сделано заключение, что анизотропия ШК УРАФ обусловлена процессом аннигиляции термализованниых позитронов с электронами внешних оболочек ионных остовов в кристаллических решетках алмазоподобных полупроводников.

Introduction

The high-momentum component (HMC) anisotropy was first observed (ARUTYUNOV et al. 1983; ARIFOV et al.) for Ge and Si during precision measurements performed in the spectral range of $\Delta\theta = 7 \div 25$ mrad of angular distribution of annihilation photons (ADAP). The present work is dedicated to the experimental and theoretical study of HMC ADAP anisotropy in "defect-free" diamond-like semiconductors. The centre of attention was the effect of HMC ADAP anisotropy in germanium and silicon and the investigation results of the temperature influence on the above mentioned effect in germanium.

1. Experimental

GaAs samples of p-type were made by the method of non-crucible zone melting (concentration of holes is $p_{300K} \approx 10^{17}$ cm^{-3}, dislocation density is $N_d \approx 10^4 \div 10^3$ cm^{-2},

¹) Contributed paper presented during the European Meeting on Positron Studies of Defects (PSD 87), March 23–27, 1987, Wernigerode, G.D.R.

material compensation degree is about 15÷20%). The samples of p-Ge and Si were obtained from materials studied before (Arifov et al.). Different specific measures such as annealing, optimization of a single crystal growth mode, etching, etc. were taken to minimize the influence of growth lattice defects, residual impurity centres with deep levels, stoichiometry and surface layer ruptures on ADAP in the process of sample preparation. The samples were oriented along the major crystallographic directions [111], [110] and [100].

The measurements were performed at room temperature in the installation with the resolution of 1 mrad which provided a long-slit geometrical scheme of gamma-quanta registration. ^{22}Na on a tantalum substrate with the activity of $\sim$ 20 mCu was used as a positron source. The count statistics in ADAP curve maximum was about $(2\div4)\cdot10^4$. While making precision measurements in the range of HMC ADAP the statistics was changed from about $(3\div5)\cdot10^4$ for the values of gamma-quanta detection angle of $\theta = 6$ mrad to about $(1.6\div3)\cdot10^3$ at $\theta = 15$ mrad. Thus, the relative statistic measurement error of $\sigma = 100\%/I^{1/2}$ (I is the count number) in this spectral range was changed from about $\sim 0.5\%$ to $\sim 2\%$ (Figures do not show the errors as in the graph scale its value is comparable with the thickness of ADAP curve).

ADAP curves were divided into components by PAACPIT program and modernized computing procedures based on this program (see, for instance, Arifov et al.; Arutyunov).

2. ADAP anisotropy in the range of $\Delta\theta = 7\div25$ mrad

Unlike the resulting curves of ADAP for Ge, Si, GaAs having the anisotropy of the same type (Arifov et al.; Arutyunov) HMC ADAP curves in dimond-like semiconductors are appreciably different for the same crystallographic directions (Fig. 1). These differences consist in the following:

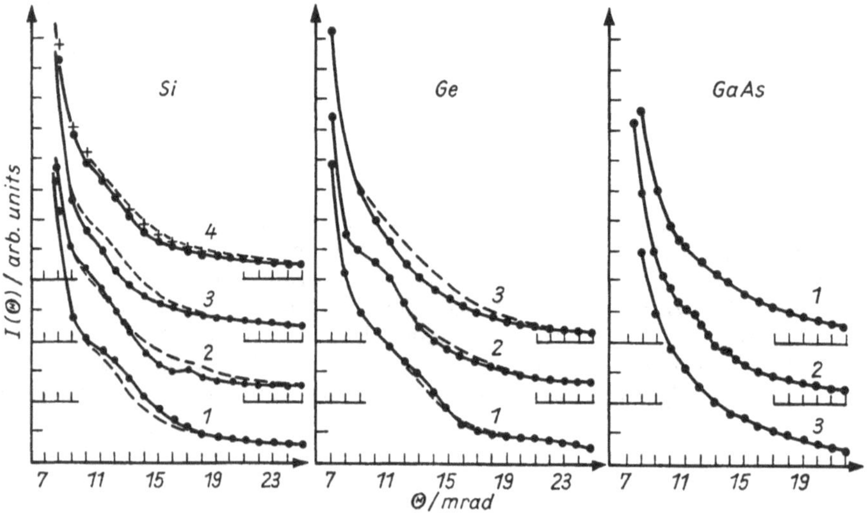

Fig. 1. Calculated (dotted lines) and experimental curves of ADAP high-momentum component for Ge, Si, and GaAs: 1 – crystallographic orientation [111], 2 – [110], 3 – [100], 4 – polycrystal silicon (the resulting curve has been obtained by ADAP spectra summing calculated for silicon oriented along crystallographic directions [111], [110] and [100]: crosses show similar results obtained by summing of experimental HMC ADAP curves. The most informative parts of the curve normalized to the value I (θ = 0 mrad) are shown

a) The results of the precision measurements demonstrate a pronounced anisotropy of HMC ADAP. For [110] direction ADAP curves contain a peculiarity ("a hump") in the angular range of $\sim 9 \div 13$ and $\sim 11 \div 13$ mrad for Ge and GaAs, respectively. Such thin structure of HMC ADAP was also observed for other dimond-like semiconductors in different crystallographic directions. Thus, for instance, "a hump" for [111] direction was found for silicon in the range of $\sim 11 \div 14$ mrad. As a whole, the smoothest curves of HMC ADAP have been obtained along [100] direction for all studied materials (due to the limited space of the paper we present only typical data; the details of the obtained results will be published elsewhere).

b) The found thin structure of HMC ADAP is connected with a certain crystallographic orientation of the crystal, for example at the transition to polycrystal Si the peculiarities of HMC ADAP disappear. In this connection it is interesting to underline that the increase of lattice atom oscillation amplitude with the temperature also results in smoothing of HMC ADAP curve for Ge [110].

It is known that the positron annihilation in a solid occurs after their thermalization. Thermalized positrons are for a certain time near the outer electron shell of the ion core. The analysis of the experimental results showed that the effect of HMC ADAP anisotropy is caused by the annihilation of thermalized positrons with core electrons. Let's take germanium and silicon for its detailed consideration.

It is known that wave functions of core electrons in Si and Ge are not symmetrical but somewhat "elongated" towards the tops of an elemental cell tetrahedron (ZAIMAN). For a central atom of the elemental tetrahedral cell the wave function of a core electron can be written as:

$$\psi_i(\boldsymbol{r}) = f(\boldsymbol{r}) + f(\boldsymbol{r} + \boldsymbol{\varrho}_i) \tag{1}$$

where $f(\boldsymbol{r})$ is a wave function of an individual ion and the vector $\boldsymbol{\varrho}_i$ defines the direction from the selected point to the nearest neighbours. According to universally accepted concepts, a positron in solids annihilates being in a thermalized state with the energy of $E = 0.03$ eV. That's why the inertion centre momentum (e^+e^-)-pair actually coincides with the electron momentum. Distribution density by $\boldsymbol{v}$ rates of the inertion centre of the annihilating (e^+e^-)-pair for an elemental tetrahedron is given by the following sum (Arutyunov et al. 1986):

$$\varphi(\boldsymbol{\beta}) = \sum_{i=1}^{4} \varphi_i(\boldsymbol{\beta}) = 4\, |f(\boldsymbol{\beta})^2 \cdot \sum_{i=1}^{4} \cos^2\left(\frac{\boldsymbol{\beta} \cdot \boldsymbol{\varrho}_i}{\lambda}\right) = 4\, |f(\boldsymbol{\beta})|^2 \cdot \eta(\boldsymbol{\beta}, \boldsymbol{\varrho}_i)\,, \tag{2}$$

where $\boldsymbol{\beta} = \boldsymbol{v}/c$ (c is the light velocity), λ is a Compton wavelength of an electron, $\varphi_i(\boldsymbol{\beta})$ is a Fourier transformation of the wave function $\psi_i(\boldsymbol{r})$. In the formula (2) the cosine defines the orientational (in respect to $\boldsymbol{\varrho}_i$) dependence of (e^+e^-)-pair momentum distribution. Within the framework of such a model the functions of ADAP $I(\theta)$ were calculated by the technique that considers the actual parameters of the experimental apparatus (ARUTYUNOV et al. 1986); to compare the calculated and the experimental ADAP we used the function:

$$S(\theta) = I(\theta)/I_p(\theta)\,, \tag{3}$$

which connects ADAP in a single and polycrystal. For the latter one $I_p(\theta)$ is defined by the function $I(\theta)$ averaged along all crystallographic directions (see ARUTYUNOV et al. 1986). Numerical values of $S(\theta)$ and $I(\theta)$ for core shells $2p^6$(Si) and $3d^{10}$(Ge) were calculated by hydrogen-like wave functions and functions of Bethe-Peierls used as $f(\boldsymbol{r})$ (BETHE, PEIERLS). These functions describe approximately correctly the behaviour of the bound electron at large distances from the nucleus appreciable for the considered model.

The detailed analysis of $S(\theta)$ function showed that its behaviour is actually independent of the type of the wave function of the core electron and it is determined by the orientational multiplier $\eta(\boldsymbol{\beta}, \boldsymbol{\varrho}_i)$ alone. The calculations demonstrated that the "structural" function $S(\theta)$ is universal for a diamond-like crystal lattice, i.e. its behaviour is rather sensitive to the values of internuclear distances and slightly depends on the binding energy of the core electron.

Figure 1 shows that the calculation results of $I(\theta)$ for HMC ADAP are in a good agreement with the experimental data, especially for [110] and [111] directions. That's why we have a good reason to assume that HMC ADAP anisotropy in dimond-like semiconductors is mainly conditioned by the process of positron annihilation with electrons of outer shells of ion cores.

3. Temperature effect on ADAP anisotropy in germanium

With the crystal temperature increase ADAP anisotropy parameters can be changed due to the growth of the oscillation amplitude of internuclear distances in the crystal and as a result of thermal equilibrium defect formation (vacancies and internode atoms) which trap positrons.

The experimental results showed that the function $I(\theta = 0, T)$ for Ge is slightly dependent on temperature in the range of $300 \leqq T \leqq 930$ K; the similar conclusion was made on the basis of the measuring data of Doppler broadening of annihilation photon energetic distribution (TANIGAWA et al.). The comparison of ADAP curves obtained at $T = 300$ K and 930 K demonstrated a slight smoothing down of distribution peculiarities in the range of $\Delta\theta = -2 \div 2$ mrad which is accompanied by a negligible narrowing of ADAP curve with the temperature increase (Fig. 2). By and large the shape of ADAP in Ge typical for [110] direction is preserved in the temperature range of $\Delta T = 300 \cdot 930$ K.

One of the typical "indicators" of positron annihilation and trapping procedure near vacancies is ADAP narrowing and $I(\theta = 0, T)$ values growth at the material temperature increase up to $\sim 0.7 T_m \div T_m$ (T_m is the melting point). Estimation of the

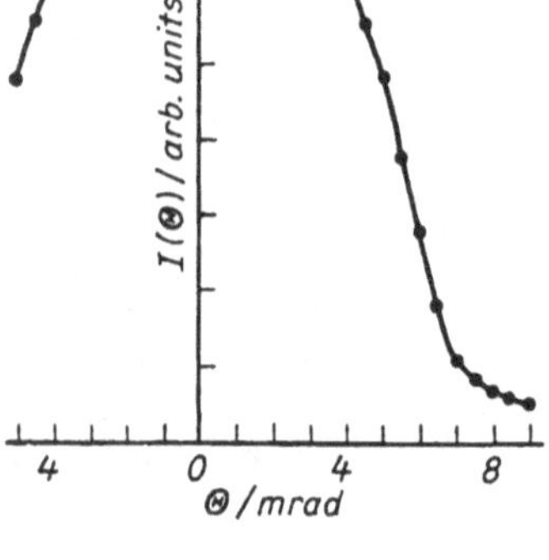

Fig. 2. ADAP curves for germanium obtained at different temperatures: small points are for $T = 930$ K (4.9 ± 0.05 mrad), big ones (upper points at the top of the curve) are for $T = 300$ K (5.05 ± 0.05 mrad). FWHM of ADAP curves normalized to I ($\theta = 0$) are given in brackets

average "number of meetings" ν of the thermalized positron during its life-time with mobile negatively charged vacancies in germanium at $T = 900 \div 1000$ K presents a numeric value $\nu = 10^3$ (vacancy migration energy during self-diffusion was assumed to be $E_m^{1V} = 0.15$ eV (TANIGAWA et al.). Thus, the obtained results indicate that positrons are not trapped by vacancies in Ge as opposed to the majority of metals.

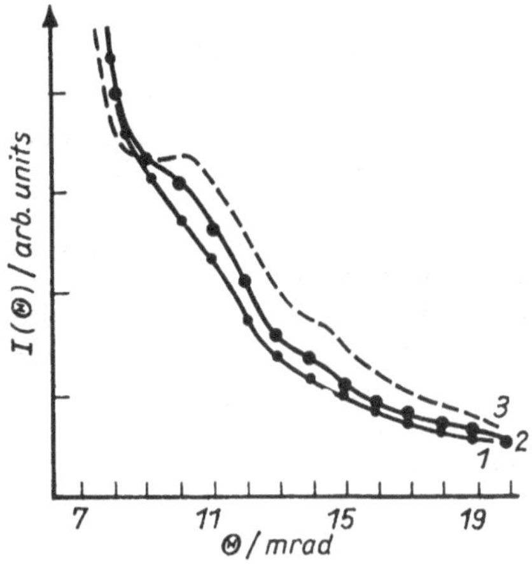

Fig. 3. Curves of ADAP high-momentum component obtained for Ge [110] at different temperatures (solid lines indicate experimental data): 1 — $T = 930$ K, 2 — $T = 300$ K, 3 — "ideal" ADAP calculated according to (7) with the help of experimental data obtained at $T = 300$ K. The curves have been normalized to the value I ($\theta = 9$ mrad); the increase of the area under the curve 1 evaluated from the experimental data in comparison to the curve 2 was $\sim 16 \div 20$ %

With the temperature growth HMC contribution to the resulting ADAP increases. In this case HMC ADAP curve becomes smoother (Fig. 3). These variations in ADAP are apparently caused by the increase in the oscillation amplitude of regular lattice atoms at the temperature growth. This fact is an additional indication of the absence of the direct effect of positron interaction with intrinsic defects of the crystalline lattice on ADAP characteristics. Actually the amplitude of internuclear distance oscillations $\boldsymbol{\varrho}_i$ being close to the equilibrium value follows the normal law of distribution and that's why the real function $\eta_T(\boldsymbol{\beta}, \boldsymbol{\varrho}_i)$ in the formula (2) is a superposition of $\eta(\boldsymbol{\beta}, \boldsymbol{\varrho}_i) = \sum_{i=1}^{4} \cos^2 [(\boldsymbol{\beta}, \boldsymbol{\varrho}_i)/\lambda]$ for different $\boldsymbol{\varrho}_i$:

$$\eta_T(\boldsymbol{\beta}, \boldsymbol{\varrho}) = \sum_{i=1}^{4} (2\pi\sigma)^{-1} \int \exp\{-(\boldsymbol{\varrho}_i - \boldsymbol{\varrho}_{i0})/\sqrt{2}\,\sigma\}^2 \cdot \eta(\boldsymbol{\beta}, \boldsymbol{\varrho}_i)\, d\boldsymbol{\varrho} \tag{4}$$

where $\sigma(T)$ is a mean square deviation of a regular atom from the equilibrium position in the lattice related to the melting point T_m of a solid by Lindeman law (ZAIMAN). $\sigma(T) \approx 0.25 \cdot \Delta_\varrho \cdot \sqrt{T/T_m}$. Substituting $\sigma(T)$ into (4) it is easy to show that $\eta_T(\boldsymbol{\beta}, \boldsymbol{\varrho})$ can be written as:

$$\eta_T(\boldsymbol{\beta}, \boldsymbol{\varrho}_i) = \sum_{i=1}^{4} \left\{\frac{1}{2}\left(1 + \exp\left[-\frac{\theta_i^2 T}{8T_m}\right] \cdot \cos 2\theta_i\right)\right\}; \tag{5}$$

$$\theta_i = \frac{\varrho \cdot \beta}{4} \cdot \sin \theta_{\varrho i} \cdot \sin \varphi_{\varrho i} = \boldsymbol{\beta} \cdot \boldsymbol{\varrho}/\lambda\,, \tag{6}$$

where $\theta_{\varrho i}$, $\varphi_{\varrho i}$ are spherical angles that determine the position of $\boldsymbol{\varrho}_i$ in the coordinate system connected with the installation (the values of the parameters in (6) are presented by ARUTYUNOV et al. 1986. At $T \to 0$ $\eta_T(\boldsymbol{\beta}, \boldsymbol{\varrho}_i)$ is transferred to $\eta(\boldsymbol{\beta}, \boldsymbol{\varrho}_i)$ where lattice atom oscillations are not taken into account.

Expressions (2) and (5) permit to calculate the function $I(\theta, T)$ of HMC ADAP in approximations considered above. To compare the experimental and theoretical data we used the following function

$$S(\theta, T) = I_{\text{exper.}}(\theta, T)/I(\theta, T = 0) \tag{7}$$

related to a single crystal ADAP obtained at the measuring temperature with "ideal" ADAP at $T = 0$ K. The function $S(\theta, T)$ is actually independent on the type of the wave function of a core electron but is determined solely by the orientational multiplier $\eta_T(\boldsymbol{\beta}, \boldsymbol{\varrho})$ and its constituent factor $\sim \exp[(-\theta_i^2/8) \cdot (T/T_m)]$. The increase in the amplitude of internuclear distances oscillations $\Delta\varrho_i$ at temperature growth results in an appreciable change of the parameters $S(\theta, T)$ and gradual smoothing of HMC ADAP curve.

Positron life-time is significantly larger than the period of thermal oscillation of lattice atoms. Due to this fact their amplitude increase with the temperature is equal to the growth of the effective volume of the crystal lattice which is occupied by ion cores. As a result the probability of positron annihilation with core electrons rises as a whole and HMC contribution to the resulting ADAP is increased. This result is in a good agreement with the experimentally observed increase in HMC contribution to ADAP for A^3B^5 semiconductors at ion core volume growth (Arifov et al.; Arutyunov). Thus, the found changes in ADAP parameters in germanium are conditioned by two reasons: "thermal" growth of HMC contribution to the resulting ADAP and smoothing down of distribution peculiarities and quite possibly the slight changes of parameters of Fermi distribution of electrons in the valent zone of germanium at temperature increase.

Conclusion

The considered results show that due to positron penetration to the volume of ion cores the core electrons in addition to valent ones participate in the annihilation process. Anisotropy of HMC ADAP in dimond-like semiconductors is mainly conditioned by positron annihilation with electrons of outer shells of ion cores. That's why the further development of positron spectroscopy based on the study of ADAP anisotropy effect can provide new information on the structure of ion cores in semiconductors.

Equilibrium vacancies do not trap positrons in germanium and due to this fact their presence in the material actually does not affect the momentum distribution of (e^+e^-)-pairs in the temperature range of $\Delta T = 300 \div 930$ K. It is probably connected with the compressive stress in the crystal lattice causing the appearence of repulsive forces between a positron and a vacancy.

References

Arifov, P., Arutyunov, N., Traschakov, V.: FTP **17** (1983) 1926
Arutyunov, N.; Ph. D. Thesis, Leningrad 1981
Arutyunov, N., Arifov, P., Traschakov, V.: Abstr. 33 All-Union Conf. on Nucl. Spectroscopy and Atomic Nucleus Structure, Moscow 1983, p. 529
Arutyunov, N., Baltenkov, A., Gilerson, V., Traschakov, V.: FTP, **20** (1986) 1753
Bethe, H., Peierls, R.: Proc. Roy. Soc. **148** (1935) 146
Tanigawa, S., Hinode, K., Nagai, R. et al.: Appl. Physics **18** (1979) 81
Zaiman, J.: Solid Theory Fundamentals, Moscow **1974**, p. 265

(Received June 16, 1987)

Authors address:

N. Arutyunov, A. Baltenkov, I. Segal
Ministry of Geology UzSSR, Middle-Asiatic Institute of Geology
and Mineral Raw Materials
T. Shevchenko str., 15
700061 Tashkent, USSR

Cryst. Res. Technol.	23	1988	2	237–241

A. V. Voitsekhovskii, A. P. Kokhanenko, A. S. Petrov,
Yu. V. Lilenko, A. D. Pogrebnyak

Siberian Physico-Technical Institute, Tomsk State University, USSR, and
Nuclear Physics Institute, Polytechnical Institute, Tomsk, USSR

Investigation of Radiation Defects in Electron Irradiated $Hg_{1-x}Cd_xTe$ Crystals Using Positron Annihilation[1])

The electron characteristics of defects in the initial and electron irradiated $Hg_{1-x}Cd_xTe$ (2–3 MeV, 10^{18} cm^{-2}, 300 K) crystals using the positron annihilation method have been investigated. The data of electric measurements are confirmed on connection of p-type conductivity with vacancy defects of metal sublattice initial crystals $Hg_{1-x}Cd_xTe$. An analysis of correlation curves of irradiated crystals has shown a possibility of formation of associations of initial defects and radiation damages of vacancy type during radiation process. The presence of narrow component on correlation curves in the region of small angles is associated with formation of positronium states localized in the region of radiation defect complex of vacancy type. Identification of positron-sensitive defects with electrically active radiation induced ones has been carried out according to the results of isochronal annealing of irradiated crystals.

Исследованы электронные свойства дефектов в исходных и облученных электронами (2—3 МэВ, 10^{18}см$^{-2}$, 300 К) кристаллах $Hg_{1-x}Cd_xTe$ методом позитронной аннигиляции. Подтверждены данные электрических измерений о связи дырочного характера проводимости с вакансионными дефектами катионной подрешетки исходных кристаллов $Hg_{1-x}Cd_xTe$. Анализ корреляционных кривых облученных кристаллов показал возможность образования в процессе облучения ассоциаций исходных дефектов и радиационных нарушений вакансионного типа. Наличие узкой компоненты на корреляционных кривых в области малых углов связывается с возникновением состояний позитрония, локализованного в области скоплений радиационных дефектов вакансионного типа. По результатам изохронного отжига облученных кристаллов проведена идентификация позитронно чувствительных дефектов с электрически активными дефектами, введенными радиацией.

1. Introduction

Identification of the defects in the $Hg_{1-x}Cd_xTe$ compounds is of principle character since the properties of the defects determine electrical and recombination characteristics of crystals (Voitsekhovskii, Lilenko). The directed variation of parameters of semiconductor $Hg_{1-x}Cd_xTe$ material using high-energy particle and ion irradiation is followed by formation of radiation defects determining the properties of irradiated crystals (Voitsekhovskii et al.). In this paper the characteristics of radiation defects in $Hg_{1-x}Cd_xTe$ crystals were investigated by the positron annihilation method after

[1]) Contributed paper presented during the European Meeting on Positron Studies of Defects (PSD 87), March 23–27, 1987, Wernigerode, GDR.

irradiating by electrons with high energy and large integral dose, and for interpreting the results not only the dose dependences of the parameters but also the data of isochronal annealing of defects were used.

2. Experimental methods

The investigations were made using $Hg_{1-x}Cd_xTe$ ($x = 0.20$), grown by the method of solid state recrystallization of p- and n-type of conductivity with concentrations of holes and electrons, respectively equaling $p = (2 \div 4) \cdot 10^{17}\ cm^{-3}$ and $n = (1 \div 2) \cdot 10^{14}\ cm^{-3}$ at $T = 77$ K. The samples of dimensions $15 \times 10 \times 0.5\ mm^3$ were irradiated by electrons of 3 MeV energy and integral fluxes up to $\Phi = 1 \cdot 10^{18}\ cm^{-2}$ using a pulse electron accelerator with pulse duration of 4 μs and pulse repetition frequency rate 0.25 kHz. The temperature of sample irradiation at average current density of 2 μA/cm² was $280 \div 300$ K and was kept constant by liquid nitrogen vapour cooling. The irradiated samples were exposed to isochronal annealing in the normal atmosphere in the temperature range of $330 \div 430$ K during 20 min. The measurement of curves of annihilation photon angular distribution (APAD) was carried out with angle resolution 0.8 mrad. The positron source was used on the basis of $^{22}Na(\beta^+, \gamma)$ with 12 mCi activity. The region of angular values investigated was ± 20 mrad. The count at the APAD curve maximum amounted to about $(9 \div 12) \cdot 10^3$ pulses, the background of random coincidences being about 1.5%. To determine the positronium states the APAD curve measurement procedure was used with the static magnetic field with the induction 1T. The APAD curve was decomposed to two components (BROUDNYI et al.) The wide component (Gaussian) and parabolic component have the following parameters: h_g, h_p — heights and Q_g — characteristic angle and Q_F — the Fermi angle. The narrow component was also presented as a Gaussian curve. The decomposition errors did not exceed the experimental error in determining the characteristic angles of components. Besides, for analyzing the experimental data the values of the ratios of areas for parabola (F_p) and Gaussian curve (F_g) to the entire area of correlation curve (F). Its width at half-height of the APAD curve $\Gamma_{1/2}$ was defined as well.

3. Experimental results and discussion

The correlation curve of initial sample of p-$Hg_{1-x}Cd_xTe$ is given in Figure 1a. At large angles θ the APAS curve broadening is observed due to the contribution of the process of positron annihilation with bound electrons. In this case, from the experimental data for initial samples it follows that for hole crystals the APAD curve is narrower and the counting rate in maximum is greater as compared with the correlation curve for crystals with electron conductivity. Curve narrowing characterizes, primarily, the large concentration defects that confirms the results of electric measurements on the presence of high concentration of vacancy defects (mainly V_{Hg}) in the p-type semiconductor determining the concentration of charge carriers (RODOT).

Electron irradiation of the samples leads to an essential variation of the type of correlation curves (Tab. 1) as at irradiation the process of positron flux redistribution according to various annihilation channels takes place. The values of θ_F in the experimental curves are in the $5.3 \div 5.8$ mrad region that is close to the calculated value 5.86. Lesser value of θ_F in the experiment indicates the predominant positron annihilation near the damage lattice, for example, vacancy defects. Minimum value of

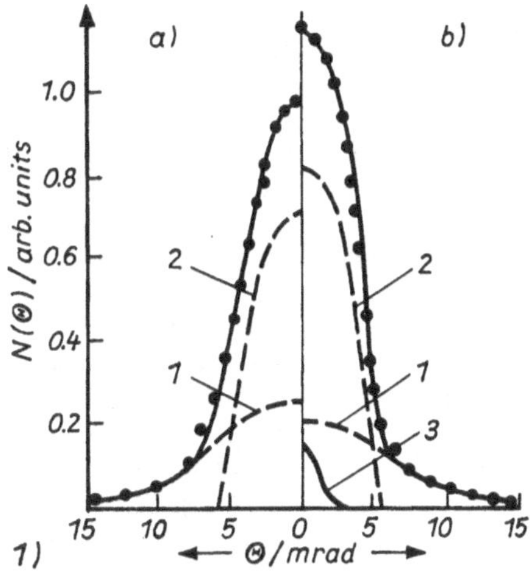

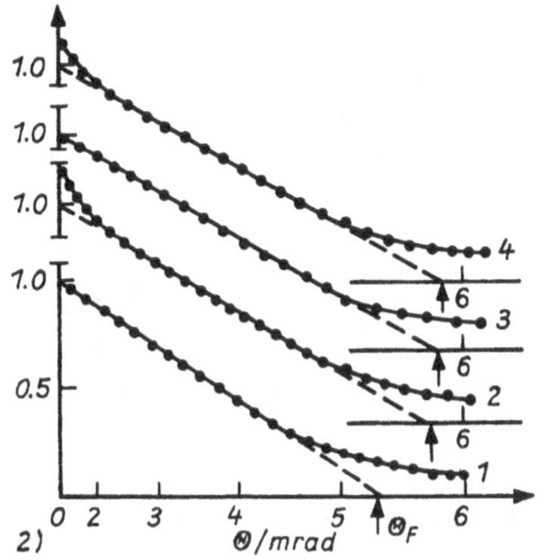

Fig. 1. Decomposition of correlation curves to the components for the sample of p-$Hg_{1-x}Cd_xTe$; initial (a) and after electron irradiation by the dose 10^{18} cm^{-2} (b) (1 – Gaussian curve, 2 – parabolic curve, 3 – narrow component)

Fig. 2. Correlation curves for p-type (1, 2) and n-type (3, 4) samples before (1, 3) and after electron irradiation by the dose 10^{18} cm^{-2} (2, 4)

θ_F belongs to the initial sample of p-type conductivity ($\theta_F = 5.33$ mrad) having high concentration of defects of acceptor type (vacancies of Hg). The θ_F increase with the irradiation increase, especially for p-type crystals, can be associated with the decrease of concentration of single vacancy defects under radiation effect. When irradiating by electrons the values of the ratio F_F/F of correlation curves varied. In this case, for the sample of p-type the ratio of the area of parabolic curve to the total curve area essentially increases with irradiation increase as compared with almost stable value of F_p/F for n-type. Therefore, so great variations of the F_p/F ratio cannot designate the equivalent increase of the electron number under the radiation effect. It is more probable that this fact denotes the positron annihilation on vacancy defects produced by irradiation. At the same time, the proximity of the F_p/F values for irradiated samples of different types of conductivity can be interpreted as a contribution excess to positron annihilation of radiation defects over the initial ones.

From Figure 1b follows that at irradiation the appearance of narrow component on the APAD curve was observed in the region of small angles. For clearer separation of narrow component in Figure 2 the APAD curves are shown as a function of θ^2.

Table 1
Parameters of APAD curves for the $Hg_{1-x}Cd_xTe$ samples before and after electron irradiation

Number of Sample	n(p)-type	Φ/cm^{-2}	θ_g/mrad	θ_F/mrad	F_p/F / arb. units	$\Gamma_{1/2}$/mrad
1	n	0	8.52	5.72	0.66	8.8
		10^{18}	8.76	5.75	0.65	8.4
2	p	0	8.12	5.33	0.56	8.6
		10^{18}	8.21	5.65	0.69	8.2

The deviation from the parabolic law of experimental dependence $N(\theta^2)$ takes place for the angles of $\theta < 2$ mrad. It is evident that the growth of the number of coincidence pulses at small angles (appearance of narrow component) results in the relative narrowing of the total APAD curve.

The presence of narrow component on the APAD curve can be due to formation of a positronium atom. For the n-type conductivity samples irradiated by electrons the application of static magnetic field leads to the narrowing of the APAD curve: the decrease of halfwidth of narrow component with the intensity increase.

In this case, the pulse distribution of positronium states ($1.0 \div 1.5$ mrad) is wider than for free positronium (0.8 mrad). This indicates that the positronium atom is in the bound state. Taking into account that the parameters of narrow components of correlation curves depend on the dimensions of the region of positronium localization, its linear dimension should be determined. The estimate of averaged size of localization according to the method (BROUDNYI et al.) gives the value of $7 \div 9$ Å that denotes the excess of dimensions of the produced vacancy complexes over the value of constant lattice of the investigated compound. It is assumed that the radiation vacancy defects and initial growth defects (V_{Hg}) are included in complexes produced during irradiation.

Figure 3 shows the behaviour of the parameter $R = (N(\theta = 0) - N_T(\theta = 0))/(N(\theta_j) - N_T(\theta_j))$ is given with annealing temperature, where $N(\theta = 0)$ and $N_T(\theta = 0)$, $N(\theta_j)$ and $N_T(\theta_j)$ are the rates of photon counting in maximum of correlation curve and at large angles (θ_j was taken equaling 9.5 mrad) after irradiation and after annealing at temperature T. The parameter R characterizes the variation of APAD curve due to redistribution of relative contribution of positrons annihilating at small and large angles θ. The value R does not depend on the defect concentration and is determined by variation of cross-sections of positrons capture by defects or by variation of the defects type (BROUDNYI et al.). As follows from the type of curves of isochronal annealing, one can separate out two stages of defect annealing being sensitive to positrons: $60 \div 100$ °C and $100 \div 160$ °C. At the first stage of annealing one can observe the increase of portion of positrons, being captured by the defects of vacancy type, i.e., "reverse" annealing that confirms the data of electrophysical measurements (VOITSEKHOVSKII et al.). Variation of the parameter R in this stage is also in agreement with the conclusion on complex formation with participation of initial and radiation defects. The second stage of annealing is characterized by restitution of the initial parameters in APAD curves, i.e., by dissociation of defects that coincides with the stage of annealing of electrically active radiation defects (VOITSEKHOVSKII et al.).

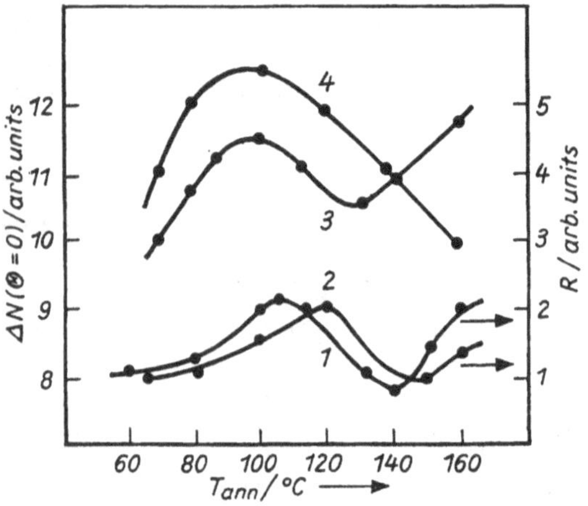

Fig. 3. Dependences of the parameters N ($\theta = 0$) and $R(\theta)$ on the temperature during isochronal annealing ($\Delta t = 20$ min) for p-type (1, 3) and n-type (2, 4) samples of electron irradiated by dose 10^{18} cm^{-2}

4. Conclusion

The presence of high concentration of vacancy defects in the hole crystals is confirmed by the method of positron annihilation. Significant variation of the correlation curves parameters with electron irradiation and subsequent annealing was revealed that was explained by the peculiarities of radiation defect formation in $H_{1-x}Cd_xTe$ crystals. The formation of the positronium atom being in the bound state and localized in the region of radiation damage of vacancy type is observed.

References

BROUDNY, V. N., ERMATOV, C. E., NURMAGAMBETOV, C. B., POGREBNYAK, A. D., TOLEBAEV, V. T.: Fizika Tverd. Tela **26** (1984) 1452

RODOT, H.: C.R. Acad. Sci. Paris **258** (1964) 6386

VOITSEKHOVSKII, A. V., LILENKO, YU. V.: phys. stat. sol. (a) **67** (1981) 381

VOITSEKHOVSKII, A. V., BROUDNYI, V. N., LILENKO, YU. V., KRIVOV, M. A., PETROV, A. S.: Sol. Stat. Commun. **31** (1979) 105

VOITSEKHOVSKII, A. V., LILENKO, YU. V., KOKHANENKO, A. P., PETROV, A. S.: Rad. Ef. **61** (1982) 79

VOITSEKHOVSKII, A. V., KOKHANENKO, A. P., PETROV, A. S.: phys. stat. sol. (a) **90** (1985) 241

(Received June 16, 1987)

Authors' address:
c/o Dr. A. V. VOITSEKHOVSKII
Siberian Physico-Technical Institute
at Tomsk State University
Revolution Square, 1
Tomsk, 634050
USSR

Book Review

I. Mertig, E. Mrosan, P. Ziesche

Multiple Scattering Theory of Point Defects in Metals: Electronic Properties

Teubner-Texte zur Physik, Band 11

BSB B. G. Teubner Verlagsgesellschaft, Leipzig 1987, 220 Seiten, 27,50 M, ISBN 3-322-00440-6

In den sechziger Jahren wurde die Berechnung der elektronischen Energiebänder zum Routineverfahren entwickelt, die Charakteristika der Bandstruktur von Metallen und die typischen Eigenschaften von Fermiflächen wurden verstanden. Aufbauend auf diesen Ergebnissen wurde in den siebziger Jahren verstärkt versucht, den Einfluß von Punktdefekten auf die elektronischen Eigenschaften zu verstehen. Die Entwicklung bezüglich des Verständnisses des Einflusses von Punktdefekten auf die Fermiflächenparameter und auf die Transporteigenschaften wurde durch die Autoren maßgeblich mitbestimmt. In der Abhandlung werden zum einen die theoretischen Methoden beschrieben, die zur Berechnung der Elektronenstruktur von Punktdefekten herangezogen werden, zum anderen werden die Zusammenhänge zwischen bestimmten Meßgrößen und der Elektronenstruktur besprochen. Weiterhin werden numerische Ergebnisse und experimentelle Daten zusammengestellt und diskutiert.

In den Kapiteln 2 bis 4 über die theoretischen Grundlagen nimmt die quantenmechanische Streutheorie den wichtigsten Platz ein. Basierend auf der Dichtefunktionaltheorie und auf dem muffin-tin-Konzept werden die charakteristischen Begriffe und Zusammenhänge der Streuung von Elektronen dargestellt. Die Betrachtungen basieren auf der Berechnung der Greenschen Funktion des idealen Wirtskristalls mittels der KKR-Methode und der Lösung der Dysongleichung für die Greensche Funktion des Kristalls mit Punktdefekt.

Die Darlegungen sind so detailliert, daß sie als Ausgangspunkt für den Start numerischer Rechnungen dienen können. Dieser Gesichtspunkt wird durch ein Kapitel über numerische Aspekte gefördert, in dem insbesonderer Fragen der Ausführung von Integrationen über die Brillouinzone besprochen werden.

Als Meßgrößen werden de-Haas-van-Alphen-Effekt, Restwiderstand, Magnetowiderstand und Hall-Koeffizient behandelt. Die theoretischen Erläuterungen zu diesen Meßgrößen starten vom Bild des gestörten Elektronengases und führen zu den gestörten Blochzuständen. Als den Einfluß des Punktdefektes bestimmende Größen können effektive Streuphasen eingeführt werden. Durch deren Verwendung ergibt sich in den Formeln eine weitgehende Analogie zu dem Fall des gestörten Elektronengases. Bei der Berechnung des Magnetowiderstandes wird insbesondere auf die Methode der Fermiflächen-Harmonischen eingegangen.

Im letzten Kapitel werden berechnete und gemessene Werte der genannten Größen zusammengestellt. Diese Daten stammen im wesentlichen aus Arbeiten der letzten fünf bis zehn Jahre. Damit schließt sich der Kreis zu den Bemerkungen am Beginn der Rezension.

In diesem Teubner-Text werden Forschungsergebnisse der letzten Jahre verarbeitet. Das Buch ist zum Einarbeiten in das Gebiet geeignet, es gibt aber auch detaillierte Auskünfte und Literaturhinweise auf die bisher erzielten Ergebnisse.

P. Rennert

Cryst. Res. Technol.	23	1988	2	243–246

A. SEN GUPTA, P. MOSER, C. CORBEL, P. HAUTOJÄRVI, P. SEN

CEN-Grenoble, DRF/SPh/MP, and
CEN-Saclay, INSTN, BP n° 2, and
Lab. of Physics, Helsinki University of Technology, and
Cyclic Acc. Division, Saha Institute of Nuclear Physics, Calcutta

Annealing of Electron Irradiated GaP Studied by Positron Lifetime Technique[1])

GaP single crystals were irradiated by 3 MeV electrons at 20 K to a dose of 4×10^{18} e/cm². An isochronal annealing in temperature region 77 ÷ 650 K followed the irradiation. Positron lifetime measurement indicated the presence of irradiation-induced vacancies in Ga sublattice. The vacancies disappeared at two stages observed at temperatures 200 ÷ 300 K and 450 ÷ 550 K.

1. Introduction

In recent years, increasing attention has been paid to study the behaviour of the impurities and the point defects which introduce deep levels in III–V compound semiconductors (BARDELEBEN). Most of these defects are still now unidentified, since they are complex in nature resulting from the interaction of the intrinsic defects (vacancies, interstitials, antisites) with various impurities present initially in the material.

Positron annihilation is an experimental technique to detect vacancies also in semiconductors. Several investigations have already been performed in III–V compounds. Microscopic identification of native vacancies (DLUBEK et al. 1986a, b; STUCKY et al.; DANNEFAER, KERR) as well as the annealing behaviour of radiation induced defects (DLUBEK 1986b, 1987; HAUTOJÄRVI et al.) have been studied.

In this paper we present results on positron trapping in vacancy defects in GaP introduced by electron irradiation. The characterization of vacancies and their annealing behaviour are also discussed.

2. Experimental details

The samples for the present study were undoped, n-type GaP single crystals with an initial carrier concentration of 2×10^{16} cm^{-3}. Prior to the irradiation the specimens were annealed at 900 K for 2 hours in Argon atmosphere in order to reduce the positron trapping in as-grown defects. This annealing made the specimens semi-insulating.

The specimens of dimensions $5 \times 3 \times 0.5$ mm³ were irradiated by 3 MeV electrons from a Van-de-Graaff accelerator at CEN-Grenoble to a dose of 4×10^{18} e$^-$/cm². The samples were placed in a liquid hydrogen cryostat for maintaining the temperature at 20 K during irradiation.

After irradiation, the specimens were transferred at 77 K in a liquid N_2 cryostat and an isochronal annealing treatment was performed in the temperature region 77–650 K at an interval of 25 K/30 min. The stability of the set temperature was better than ±0.3 K.

[1]) Contributed paper presented during the European Meeting on Positron Studies of Defects (PSD 87), March 23–27, 1987, Wernigerode, GDR.

Positron lifetime measurements were performed by using a ^{22}Na positron source evaporated on a thin Ni foil of thickness 1.0 μm and all measurements were carried out at 77 K following an in-situ annealing at each set temperature.

A conventional slow-fast coincidence system of a resolution of 280 ps (FWHM) for Co-60 prompt source at the experimental positron window setting was used for positron lifetime measurements. A BaF_2 scintillator was used in the STOP channel to increase the counting rate.

After proper source corrections all lifetime spectra were analyzed by using only one exponential component. The variance of the fit was maintained within 1, and there was no need for a second lifetime component.

Results and discussion

Pre-irradiation

The positron lifetime value obtained before irradiation was 225 ± 2 ps. This value corresponds to the lifetime values obtained also after post-irradiation isochronal annealing above 550 K. Hence it can be assigned to the positron lifetime in the bulk material (τ_b) and it agrees with the value obtained earlier by DLUBEK et al. (1986b).

Moreover, the present value ($\tau_b = 225 \pm 2$ ps) lies in between the experimental bulk lifetime values for Si ($\tau_b = 220$ ps) (DANNEFAER, MASCHER) and Ge ($\tau_b = 228$) (MOSER et al.).

Post-irradiation

Figure 1 shows the isochronal annealing curve for electron irradiated GaP in the temperature region 77 ÷ 650 K. The average lifetime $\bar{\tau}$ has been plotted against the annealing temperature. A single component fit of the lifetime data gives a lifetime value of 240 ± 2 ps in irradiated GaP at 77 K. This is high compared to the bulk lifetime τ_b showing positron trapping at vacancy defects.

The annealing is rather continuous, but there are indications for two stages at temperatures 200 ÷ 300 K and 450 ÷ 550 K, respectively. Beyond 550 K, the average lifetime reaches the bulk value meaning the disappearence of positron trapping centers.

As positrons are positively charged they can be attracted by neutral or negatively charged vacancies, whereas positively charged vacancies repel them. Hence, positrons

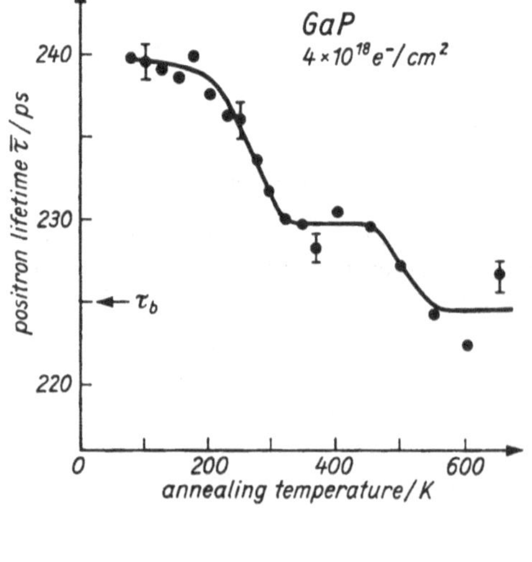

Fig. 1. Recovery of average positron lifetime as a function of annealing temperature

are sensitive to the change in the vacancy charge state which is due to the shift in the Fermi level positions.

In analogy with GaAs, we assume that in GaP the vacancy at Ga-site (V_{Ga}) is of acceptor type and can exist in neutral or negatively charged states whereas the vacancy at P-site (V_P) is of donor type and can be in neutral, negative or positive states, depending on the position of the Fermi level (SHARMA, MADHUKAR; MOONEY, KENNEDY). Positron trapping is therefore depending on the charge state of V_p, as a positively charged P-vacancy is unable to trap a positron. In n-type materials V_p is neutral or negatively charged, whereas in p-type or in semi-insulating materials V_p becomes positively charged.

After heavy irradiation at a high dose of $\Phi = 4 \times 10^{18}$ e/cm^2, the specimen becomes semi-insulating independent on its original state, and the Fermi level lies in the midgap position. Therefore, after irradiation the trapping of positrons must occur due to defects involving Ga-vacancies, as only Ga-vacancy (V_{Ga}) remains neutral or negatively charged even in semi-insulating materials.

The results in Figure 1 and the average lifetime value $\tau = 240 \pm 2$ ps in the as-irradiated state are based on one exponential fits. We are, however, not sure whether after irradiation positron trapping in Ga-vacancies is in saturation. DLUBEK et al. (1986b) found in as-grown GaP crystals a second lifetime component $\tau_2 = 290$ ps, which they attributed to P-vacancies. After neutron irradiation the same authors found $\tau = 250$ ps which was attributed to Ga-vacancies. The latter value is comparable to our value of 240 ps. Since neutron irradiation can create also some multiple vacancies we may conclude that positron lifetime in a Ga-vacancy in GaP is $\tau(V_{Ga}) = 240$ to 250 ps. For comparison, in GaAs the lifetime $\tau(V_{Ga}) = 260$ ps, and $\tau_b = 230$ ps (HAUTOJÄRVI et al.). Thus in both GaP and GaAs $\tau(V_{Ga}) \cong 1.1\, \tau_b$.

The observation of the annealing behaviour in the present investigation shows a recovery stage between 200 and 300 K. This coincides with the annealing between 200 to 350 K that has been observed in similar experiments in e^--irradiated GaAs and that has been assigned to Ga-vacancies, too (HAUTOJÄRVI et al.). The recovery stages found after room temperature neutron irradiations are above 600 K (DLUBEK et al. 1986b), where our sample has already fully recovered. BARNES has found annealing stages at 200 K, 300 K, 600 K and 800 K by electrical measurements in GaP after low temperature neutron irradiation. His first three stages overlap with the temperature region, where we see the recovery of Ga-vacancies.

In conclusion, we see the recovery of Ga vacancy defects probably in two stages between 200 and 300 K and 450 and 550 K. However, we are not able to say whether the defects are isolated Ga vacancies or associated to some other defects.

References

BARDELEBEN, J. H. von (Editor): Defects in Semiconductors, Aedermannsdorf 1986, Mat. Sci. Forum 10–20

BARNES, C. E.: Appl. Phys. Lett. 20 (1972) 110

DANNEFAER, S., KERR, D.: J. Appl. Physics 60 (1986) 591

DANNEFAER, S., MASCHER, P., KERR, D.: Phys. Rev. Lett. 56 (1986) 2195

DLUBEK, G., BRÜMMER, O., PLAZAOLA, F., HAUTOJÄRVI, P.: J. Phys. C. 19 (1986) 331

DLUBEK, G., BRÜMMER, O., POLITY, A.: Appl. Phys. Lett. 49 (1986) 385 b)

DLUBEK, G., KRAUSE, R., BRÜMMER, O., TITTES, J.: Appl. Phys. A 42 (1987) 125

HAUTOJÄRVI, P., MOSER, P., STUCKY, M., CORBEL, C.: Appl. Phys. Lett. 48 (1986) 809

MOONEY, P. M., KENNEDY, T. A.: J. Phys. C 17 (1984) 6277

MOSER, P., PAUTRAT, J. L., CORBEL, C., HAUTOJÄRVI, P.: Proc. of 7th Int. Conf. on Positron Annihilation. Eds. P. C. JAIN, R. M. SINGRU, K. P. GOPINATHAN, Singapore 1985, pp. 733–735

Sharma, S. Das, Madhukar, A.: Phys. Rev. B **24** (1981) 2051

Stucky, M., Paulin, R., Geffroy, B., Corbel, C., Suskij, J.: Proc. of 7th Int. Conf. on Positron Annihilation, eds P. C. Jain, R. M. Singru, K. P. Gopinathan, Singapore 1985, pp. 714–716

(Received June 16, 1987)

Authors' addresses:

A. Sen Gupta and P. Moser
CEN-Grenoble, DRF/SPh/MP,
85 X — 38041 Grenoble Cedex
France

C. Corbel
CEN-Saclay, Instn
BP n°2
91191 Gif sur Yvette
France

P. Hautojärvi
Laboratory ot Physics
Helsinki University of Technology
02150 Espoo
Finland

P. Sen
Cyclic Accelerator Division
Saha Institute of Nuclear Physics
92 A.P.C. Road
Calcutta 700009
India

Cryst. Res. Technol.	23	1988	2	247–251

P. Mascher, D. Kerr, S. Dannefaer

Department of Physics, University of Winnipeg
Winnipeg, Manitoba, Canada

Low Temperature Investigations of e⁻-Irradiated GaAs[1])

Positron lifetime measurements have been performed on electron irradiated chromium doped semi-insulating GaAs as a function of annealing between 139 and 908 K. Trapping at negatively charged Ga_{As} defects and by a vacancy mixture dominated by trivacancies took place in the as-irradiated state. Upon annealing around 210 K divacancies were formed from the trivacancies as caused by migrating arsenic interstitials. Complete trapping prevailed up to 530 K but decreased then rapidly at higher temperatures due to the migration of divacancies which also remove the Ga_{As} defects.

Positronenlebensdauern wurden in elektronenbestrahltem Cr-dotiertem „semi-insulating" GaAs als Funktion von isochroner thermischer Behandlung zwischen 139 und 908 K gemessen. Nach der Elektronenbestrahlung sind sämtliche Positronen entweder durch negativ geladene Ga(As)-Antistrukturatome oder durch eine Mischung von Dreifach- und Doppelleerstellen eingefangen. — Bei 210 K findet ein Übergang von Dreifach- zu Doppelleerstellen statt. Danach wurde ein rapides Abnehmen der totalen Einfangrate oberhalb 530 K festgestellt. Dies wird durch die Mobilisierung der Doppelleerstellen und die damit verbundene Beseitigung des starken Einfangs der Positronen durch die Ga(As)-Defekte verursacht.

Introduction

GaAs is rapidly becoming a more and more important material in semiconductor technology. From the vast experience with silicon it is known that point defects created during growth and/or by interaction with impurities play an important role in the performance of devices made from the basic silicon wafers. The same holds for GaAs as indicated by the intensive investigation of defect properties in GaAs by various techniques (von Bardeleben). Despite the obvious merits of positron liftime studies of defects in metals (see for example ref. (Jain et al.)) only very few lifetime investigations on defects in GaAs have been published over the years. (Cheng et al.; Dannefaer; Kerr et al.; Dannefaer et al. 1984; Dlubek et al.; Gupta et al.; Hautojärvi et al.; Dannefaer, Kerr; Stucky et al.)

The results presented here on electron irradiated Cr-doped GaAs are the first in a series of systematic investigations of different doping types and levels.

Experimental

The samples were cut from a Cr-doped semi-insulating wafer ($\varrho \cong 2 \times 10^7\ \Omega$ cm) purchased from Cambridge Instruments, England. The samples were irradiated with 28 MeV electrons to a dose of $5 \times 10^{17}\ e^-\ cm^{-2}$ and the sample temperature during irradiation was about 130 K.

[1]) Contributed paper presented during the European Meeting on Positron Studies of Defects (PSD 87), March 23–27, 1987, Wernigerode, GDR.

The first part of the annealing sequence was done with the samples mounted in an evacuated cryostat providing a temperature range from 20 to 360 K. The annealing time was 1 hour, and the measuring temperature was 100 K. Occasionally a measuring temperature of 30 K was employed to examine possible temperature dependencies. For annealing temperatures above 360 K the measurements were done outside the cryostat and at room temperature.

The lifetime spectra were obtained with a fast-fast lifetime system having a time resolution of 250 ps FWHM. At least 6.5×10^6 counts per spectrum were accumulated over a period of about 22 hours. A foil-source-arrangement (13.7 μCi^{22}NaCl enclosed by a 1 mg cm^{-2} Al-foil) was used for the low-temperature measurements. For the measurements at room temperature we used a directly deposited ^{22}NaCl-source of 3.4 μCi. A comparison of the results of room temperature measurements with both sources lead to a two-term source-correction of 251 ps with 1.6% for the Al-foil and 492 ps with 1.1% for the NaCl-contribution.

The lifetime spectra were analyzed by POSITRONFIT (KIRKEGAARD et al.) and decomposed into three components, the longest of which, however, was a constant term of about 1.5 ns with 0.3%. This term could easily be separated from the actual spectra leaving two significant terms from the GaAs itself. A quest for one more significant lifetime was unsuccessful throughout the whole range of investigation.

Results and discussion

In Figure 1 we show the lifetimes τ_1 and τ_2 and the intensity of the τ_2-component, I_2, as a function of isochronal annealing. In addition, a lifetime "τ_B" is calculated according to the simple trapping model (BERTOLACCINI et al.). It should be the lifetime of positrons in the bulk provided the simple trapping model is satisfied.

In the as-irradiated state up to an annealing temperature of $T_A = 192$ K we find two lifetimes, $\tau_1 = 215$ ps and $\tau_2 = 315$ ps with a constant intensity ratio I_2/I_1 of

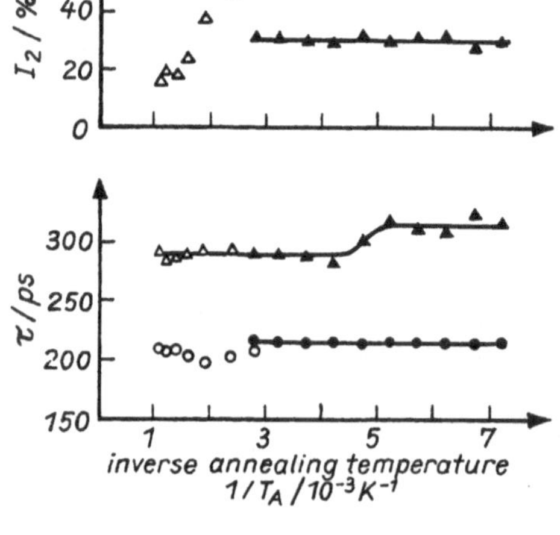

Fig. 1. Lifetime results for e-irradiated Cr–GaAs as a function of annealing temperature. Measuring temperatures are 100 K (●, ▲) and 298 K (○, △). The "bulk" lifetime is calculated from the simple trapping model

0.43. At $T_A = 212$ K the second lifetime component shows an unambiguous decrease to 290 ps without affecting either the value of τ_1 or the intensity-ratio of the two components. No measurable temperature dependence could be found when comparing data taken at 30 or at 100 K.

The fact that (a) the values of "τ_B" up to at least 420 K are unreasonably high and (b) the first (shorter) lifetime component is not affected by the significant change in τ_2 at 212 K lead us to the conclusion that we are dealing with full trapping up to $T_A = 420$ K. Therefore, **both** lifetimes have to be associated with defects in the sample rather than designating τ_1 as the usual modified bulk lifetime.

An annealing experiment on a sample of unirradiated Cr-doped GaAs from the same wafer showed that the main defect in the as-grown state is the $V_{As}Cr_{Ga}$-complex, giving a lifetime $\tau = 265$ ps. This is in very good agreement with the lifetimes determined for other vacancy-impurity pairs (Dannefaer, Kerr). By these authors also a lifetime level of 290 ps was found in undoped GaAs and has been identified to be the response from $V_{As}V_{Ga}$ divacancies. From a plot of the lifetime values for the bulk, the monovacancy-complex and the divacancy versus the multiplicity of the involved vacancy, one would expect a value of about 330 ps for the trivacancy. The value found here of 315 ± 5 ps is close, but could be an indication for not having a "pure" trivacancy response.

The following model then emerges to explain the initial stages of the induced radiation damage and its annealing. Due to the high energy of the incident electrons not only primary defects such as simple Frenkel pairs are formed but most of the electrons will cause more than one displacement of an atom on a lattice site. The displacement cascade will contain vacancies, most likely in multiple form, interstitials of both types and probably antisites. The very first step of defect removal obviously is the disappearance of the monovacancy response already in the as-irradiated state. The most likely process to achieve that is the "annihilation" of the vacancy by an interstitial. Since the charge state of the $V_{As}Cr_{Ga}$-complex is neutral (Kikoin et al.) there is no restriction as to which interstitial could be responsible. However, we will show later that the Ga interstitial, Ga_I, has to be the species of choice. The $V_{As}Cr_{Ga}$ may well be annihilated by migrating Ga-interstitials forming $Ga_{As}Cr_{Ga}$ pairs. Other possibilities for the Ga interstitials are formation of $Ga_{As} + As_I$ pairs (if their energy is high enough) or reduction of $V_{Ga} \cdot V_{As} \cdot V_{Ga}$ trivacancies to $V_{Ga} \cdot V_{As}$ divacancies. From this the main components of the defect structure we are dealing with in the as-irradiated state are: trivacancies (both $V_{Ga} \cdot V_{As} \cdot V_{Ga}$ and $V_{As} \cdot V_{Ga} \cdot V_{As}$), divacancies ($V_{As}V_{Ga}$), As-interstitials and antisites of both types. Calculations of the energy levels of different charge states of the involved defects (Baraff, Schlüter) show that only the $V_{Ga} \cdot V_{As} \cdot V_{Ga}$ trivacancy, the divacancy and the Ga_{As} antisite are potential positron traps. We therefore associate the shorter lifetime component, τ_1, with positrons localized by the Coulomb field of the negatively charged Ga_{As}-antisite (but still in a very bulk-like state) and the longer component, τ_2 with the remaining positrons being trapped by di- and trivacancies.

Above 200 K the As interstitials become mobile and — being positively charged (Baraff, Schlüter) — find the $V_{Ga} \cdot V_{As} \cdot V_{Ga}$ trivacancies to reduce them to divacancies. There are no conclusive data on the migration of As_I available. As a matter of fact published migration temperatures vary from below 100 K (Newman) to above room temperature (Pons, Bourgoin).

The next very important step as far as annealing goes is the obvious effect of the increase in measuring temperature from 100 to 300 K after annealing at 360 K for 1 hr. As one can see in Figure 1 the apparent intensity of the divacancy component increases by about 50% and τ_1 drops to 209 ps. Since there was no observable temperature dependence between 30 and 100 K we safely can rule out Fermi-level effects and suggest that we are dealing with the transfer of positrons from their quasi-localized state around the Ga_{As} antisites into the divacancies. This picture is very similar to common detrapping models (PAGH et al.) and accounts also for the observed decrease in τ_1.

Above $T_A = 420$ K we start to lose the divacancy response, a result we also found in several other GaAs samples (DANNEFAER et al. 1987). As a consequence the modified bulk-lifetime (which has been virtually zero over the range of full trapping) starts to increase and mixes into the still existing 209 ps component, thereby causing an apparent decrease in τ_1. From $T_A = 633$ K we are finally dealing with results satisfying the two-state trapping model as is indicated by the constant value of $\tau_B = 221$ ps (see Fig. 1).

In Figure 2 is shown the fraction of trapped positrons as a function of annealing temperature. The steep drop in the total fraction above 533 K and the validity of the two-state trapping model above 633 K shows that in this range the Ga_{As} antisites must have disappeared. This has also been found by COLLINS et al. from far-infrared spectroscopy. Our data strongly suggest that this disappearance is caused by the migrating divacancies rather than by migration of the Ga_{As} antisites themselves.

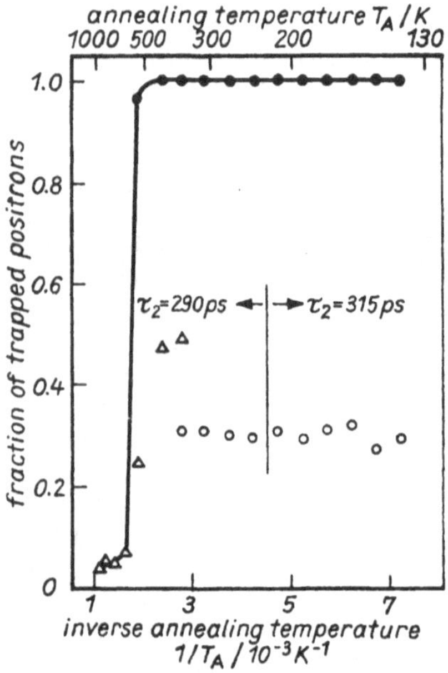

Fig. 2. Total fraction of trapped positrons (●) and fraction of positrons trapped by vacancies (○, △)

Conclusions

The results of positron lifetime measurements as a function of isochronal annealing on Cr-doped GaAs irradiated with 28 MeV electrons at 130 K can be summarized as follows:

1) Positrons are fully trapped in the as-irradiated state. The trapping centres are Ga_{As} antisites giving a lifetime of 215 ps. Vacancies are predominantly in a 3-vacancy agglomerate.

2) At 212 K As interstitials become mobile and reduce the trivacancies to divacancies. These defects show the previously found lifetime of 290 ps.

3) An increase of the measuring temperature from 100 to 300 K leads to a significant redistribution of the positrons from the antisite-state to the divacancy-traps.
4) Above $T_A = 420$ K the divacancies become mobile and destroy subsequently the Ga_{As} trapping sites. Above 633 K no more influence of Ga_{As}-antisites could be detected.

The authors would like to thank the staff of the Saskatchewan Accelerator Laboratory, Saskatoon, Canada, for their assistance during the electron irradiation. This work was supported by the Natural Sciences and Engineering Research Council of Canada.

References

BARAFF, G. A., SCHLÜTER, M.: Phys. Rev. Lett. **55** (1985) 1327
BARDELEBEN, H. J. von (ed.): Proceedings of the 14th Internat. Conf. on Defects in Semiconductors, Bay Village, Ottio 1986
BERTOLACCINI, M., BISI, A., GAMBARINI, G., ZAPPA, L.: J. Phys. C: Solid State Phys. **19** (1971) 734
CHENG, L. J., KARINS, J. P., CORBETT, J. W., KIMERLING, L. C.: J. Appl. Physics **50** (1979) 2962
COLLINS, J. D., GLEDHILL, G. A., NEWMAN, R. C.: Proceedings of the 14th Internat. Conf. on Defects in Semiconductors, 1986, p. 1081
DANNEFAER, S.: J. Phys. C: Solid State Phys. **15** (1982) 599
DANNEFAER, S., HOGG, B., KERR, D.: Phys. Rev. B **30** (1984) 3355
DANNEFAER, S., KERR, D.: J. Appl. Physics **60** (1986) 591
DANNEFAER, S., MASCHER, P., KERR, D.: in: Proc. of the 2nd Internat. Symp. on Defect Recognition and Image Processing in III-V Compounds, Monterey, New York 1987, p. 313
DLUBEK, G., BRÜMMER, O., PLAZAOLA, F., HAUTOJÄRVI, P.: J. Phys. C: Solid State Phys. **19** (1986) 331
GUPTA, A. Sen, NAIDU, S. V., SEN, P.: A **40** (1986) 95
HAUTOJÄRVI, P., MOSER, P., STUCKY, M., CORBEL, C., PLAZAOLA, F.: Appl. Physics. Appl. Phys. Lett. **48** (1986) 809
JAIN, P. C., SINGRU, R. M., GOPINATHAN, K. P. (eds): Positron Annihilation, Singapore 1985
KERR, D. P., KUPCA, S., HOGG, B. G.: Phys. Lett. **88A** (1982) 429
KIKOIN, K., LYUK, P., PERVOVA, L., VANEM, R.: J. Phys. C: Solid State Phys. **19** (1986) 4561
KIRKEGAARD, P., ELDRUP, M., MOGENSEN, M., PEDERSEN, N. J.: Computer Phys. Commun. **23** (1981) 307
NEWMAN, R. C.: in: Proc. of the 13th Internat. Conf. on Defects in Semiconductors, eds L. C. KIMERLING, J. M. PARSEY, Jr., Warrendale 1985, p. 87
PAGH, B., HANSEN, H. E., NIELSEN, B., TRUMPY, G., PETERSEN, K.: Appl. Phys. A **33** (1984) 255
PONS, D., BOURGOIN, J. C.: J. Phys. C: Solid State Phys. **18** (1985) 3839
STUCKY, M., CORBEL, C., GEFFROY, B., MOSER, P., HAUTOJÄRVI, P.: in: Proc. of the 14th Internat. Conf. on Defects in Semiconductors, ed. H. J. VON BARDELEBEN, Bay Village, Ottio 1986, p. 265

(Reveived June 16, 1987)

Authors' address:
P. MASCHER, D. KERR, S. DANNEFAER
Department of Physics, University of Winnipeg
Winnipeg, Manitoba R3B 2E9, Canada

Cryst. Res. Technol.	23	1988	2	253—258

G. DLUBEK, O. BRÜMMER, A. POLITY

Sektion Physik, Martin-Luther-Universität Halle — Wittenberg

Vacancy Defects in as-grown and n°-irradiated GaP and $GaAs_{1-x}P_x$ Studied by Positrons[1])

Measurements of positron lifetime $\bar{\tau}$ and of the shape parameter S of the Doppler-broadened annihilation line are used to study bulk and defect properties in GaP and $GaAs_{1-x}P_x$. $\bar{\tau}$ and S decrease linearly with the composition x of $GaAs_{1-x}P_x$ layers. A second lifetime component ($\tau_2 = 290$ ps) observed in as-grown GaP is attributed to stoichiometric P vacancies. After neutron irradiation of GaP and $GaAs_{0.13}P_{0.87}$ positrons are trapped by Ga vacancies ($\tau_2 = 250$ ps). These vacancies anneal out in two stages at 200—550 °C and 550—800 °C.

Messungen der Positronenlebensdauer $\bar{\tau}$ und des Formparameters S der Doppler-verbreiterten Annihilationslinie werden benutzt, um Volumen- und Defekteigenschaften von GaP und $GaAs_{x-1}P_x$ zu untersuchen. $\bar{\tau}$ und S verringern sich linear mit der Komposition x von $GaAs_{x-1}P_x$-Schichten. Eine zweite Lebensdauerkomponente ($\tau_2 = 290$ ps), welche in GaP beobachtet wird, läßt sich auf stöchiometrische P-Leerstellen zurückführen. Nach Neutronenbestrahlung von GaP und $GaAs_{0.13}P_{0.87}$ werden die Positronen durch Ga-Leerstellen ($\tau_2 = 250$ ps) eingefangen. Diese Leerstellen heilen in zwei Stufen bei 200—550 °C und 550—800 °C.⁺

1. Introduction

The microscopic identification and the physics of native point defects in III—V compound semi-conductors is far from complete. P_{Ga} antisite defects and Ga vacancies have been identified in electron spin resonance (ESR) spectra of electron-irradiated GaP. In as-grown GaP only P_{As} defects with a maximum concentration of 4×10^{16} cm^3 were detected (KAUFMANN, KENNEDY; MOONEY et al. 1983, 1984). From thermodynamical calculations (VAN VECHTEN), however, high equilibrium vacancy concentrations are expected in GaP. In this paper we report on a study of bulk and defect properties of GaP and $GaAs_{1-x}P_x$ by utilizing positron annihilation technique.

2. Experimental

For measuring the positron lifetime τ a conventional fast-slow coincidence device with a time resolution of 280 ps (full width at half-maximum) was used. A Na-22 source of 10 μCi, deposited and sealed between two 1 μm thick nickel foils, was sandwiched by two identical sample pieces (8 × 8 mm²) cut from the same wafer. For each lifetime spectrum about 10^6 pulses were collected. After source correction the spectra were analysed using one or two exponential components. From the lifetimes τ_i and the relative intensities I_i the average positron lifetime $\bar{\tau} = I_1\tau_1 + I_2\tau_2$ was also calculated. For studying the effect of isochronal annealing on point defects measurements of the Doppler-broadened annihilation line were

[1]) Contributed paper presented during the European Meeting on Positron Studies of Defects (PSD 87), March 23—27, 1987, Wernigerode, GDR.

utilized. For the line shape measurements we used a Ge(Li) detector with an energy resolution of 1.1 keV at 514 keV. To describe the line shape we calculated the so-called S-parameter defined as the relative area of a fixed central region of the annihilation line. Isochronal (0.5 hours) annealing of samples was performed in an open quartz ampoula under argon gas.

3. Composition dependence of τ and S in $GaAs_{1-x}P_x$

Figure 1 shows the variation of the average positron lifetime and of the S-parameter in $GaAs_{1-x}P_x$ as a function of the composition x. The ternary compounds are N-doped and undoped layers of 30 μm thickness grown on GaAs ($x \leqq 0.40$) or GaP ($x \geqq 0.60$) wafers of (100) orientation with the VPE and LPE technique, respectively. Between a layer of composition x and the substrate there is a transition region of 30–40 μm thickness in which the composition changes continuously from x to 0 or 1, respectively. About 70% of implanted positrons are stopped in the ternary layer, the effective thickness of which is estimated to 50 μm. Positrons penetrating the layer annihilate in the substrate below the layer.

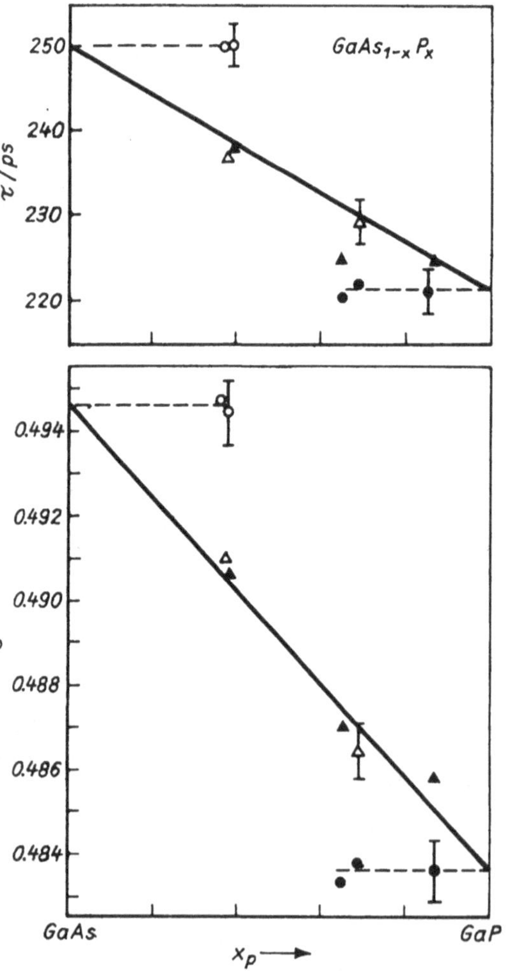

Fig. 1. Average positron lifetime $\bar{\tau}$ and S parameter of Dopplerbroadened annihilation line in $GaAs_{1-x}P_x$ as a function of the composition x. △ N-doped GaAsP, ▲ undoped GaAsP, ● GaP substrate, ○ GaAs substrate

As seen in Figure 1 $\bar{\tau}$ and S decrease linearly with the P content of the ternary compound. Nitrogen doping seems not to effect $\bar{\tau}$ and S. The annihilation parameters of the substrates were measured on the sample surface opposite to the corresponding layer. The lifetime of GaP substrate (S-doped, $n = 2-4 \times 10^{17}$ cm^{-3}) corresponds to the bulk lifetime of GaP $\tau_b = 223 \pm 2$ ps (DLUBEK et al. 1986a). The GaAs substrate (Si-doped, $n \approx 10^{18}$ cm^{-3}), however, exhibits a lifetime of 250 ps which is distinctly larger than the bulk value in GaAs, $\tau_b = 230$ ps. This situation is typical of heavily doped n-type GaAs and can be attributed to positron trapping by acceptor-type donor-Ga vacancy complexes, $Si_{Ga}V_{Ga}$ (DLUBEK et al. 1986b). The decrease of $\bar{\tau}$ and S can be attributed to the almost linear decrease of the lattice constant a with the composition x of the ternary component (GaAs: $a = 0.566$ nm, GaP: $a = 0.545$ nm). The positron lifetime is inverse proportional to the valence electron density and varies therefore as a^3. The S-parameter reflects the inverse width of the electron momentum distribution and therefore also of the Jones-Zone size which varies as $1/a$. The resulting effect is a decrease of S with decreasing a. Thus τ and S are related to the composition of a III—V semiconductor alloy and may be used for estimating the composition. This relation is however only valid if there is no positron trapping by grown-in vacancies. Further, the effect of annihilation events from the substrate must be taken into account in a proper way.

4. Vacancy defects in as-grown GaP

Positron lifetime measurements were performed on a series of polished GaP wafers of (100) orientation. The crystals grown with the LEC technique were doped with S and had a typical carrier concentration of $n = 2-4 \times 10^{17}$ cm^{-3}. Most of the samples exhibit lifetimes of 223—225 ps. In some wafers higher $\bar{\tau}$ values associated with a second lifetime component appear, $\bar{\tau} \leqq 237$ ps, $\tau_2 = 290 \pm 10$ ps, $I_2 \leqq 50\%$. This second component is evidently due to positron trapping at a vacancy-type defect. The lifetime $\tau_2 = \tau_d = 290 \pm 10$ ps is characteristic of the type of defect. The relative increase $\tau_d/\tau_b = 1.30 \pm 0.06$ ($\tau_b = 223 \pm 2$ ps) corresponds to the increase observed in GaAs and attributed to As vacancies (DLUBEK et al. 1986b). Therefore, we may conclude that the defect trapping positrons in our as-grown GaP crystals is a vacancy in the P sublattice. In n-type GaP the Fermi-level is situated near the conduction band and the P vacancies which are presumed to be of donor type (SHARMA, MADHUKAR) should be in a neutral (or possibly negative) charge state.

From the parameters I_2 or $\bar{\tau}$ we can estimate the vacancy concentration using the simple one trap trapping model. With a specific trapping rate of $\mu_d = 2 \times 10^{-9}$ cm^3 s^{-1} for monovacancies (DLUBEK et al. 1986b) a vacancy concentration of $C_d = 6 \times$ $\times 10^{17}$ cm^{-3} follows from $\bar{\tau} = 237$ ps. The appearance of such high vacancy concentration can be attributed to deviations from the stoichiometric composition of the GaP crystals. Thermoeynamical calculations by JORDAN et al. and VAN VECHTEN have shown that the equilibrium concentration of P vacancies exceeds that of Ga vacancies by one or two orders of magnitude. These calculations fit well the results of coulombmetric measurements which show that GaP is either stoichiometric or enriched with Ga. Both of these results correlate well with the interpretation that vacancies in the P sublattice trap positrons in our GaP samples. At present, however, we cannot decide whether the P vacancies are isolated or complexed with antisites or impurities. Further, the accuracity of the specific positron trapping rate used it not known. Isolated P_{Ga} antisite defects have been detected in Zn-doped GaP by ESR experiments in a concentration of 4×10^{16} cm^{-3}. In n-type GaP, however, the concentrations of P_{Ga} defects were in the 10^{15} cm^{-3} range or below (MOONEY et al. 1983, 1984).

5. Defects in neutron-irradiated GaP and $GaAs_{0.13}P_{0.87}$

A layer of $GaAs_{0.13}P_{0.87}$ grown on a GaP substrate was irradiated in the nuclear reactor of the ZfK Rossendorf with a fluence of 8×10^{18} cm^{-2} fast neutrons. To keep transmutation low, slow neutrons were shielded by cadmium. The sample temperature during irradiation did not exceed 90 °C. Three months after irradiation the residual activity of samples was low enough to be used for positron annihilation measurements. After irradiation the positron lifetime of both the layer and the substrate increases from 225 ps and 223 ps to a value of 250 ps.

Neutron irradiation generates displacement cascades containing point defects in both sublattices in the order of 10^3. Irradiated samples are semi-insulating and the Fermi level is in a mid-gap position. Then P vacancies are positively charged and repel positrons whereas the negatively charged Ga vacancy is able to localize positrons. Positron trapping by defects can be assumed to be saturated for such a high neutron fluence as 8×10^{18} cm^{-2}. Therefore, the average positron lifetime of $\bar{\tau} = 250$ ps should correspond approximately to the value typical of Ga vacancies. The corresponding increase in lifetime $\tau_d/\tau_b = 1.12$ correlates well with one which has been observed in electron and neutron irradiated GaAs and attributed to Ga vacancies (HAUTOJÄRVI et al.; DLUBEK et al. 1986c).

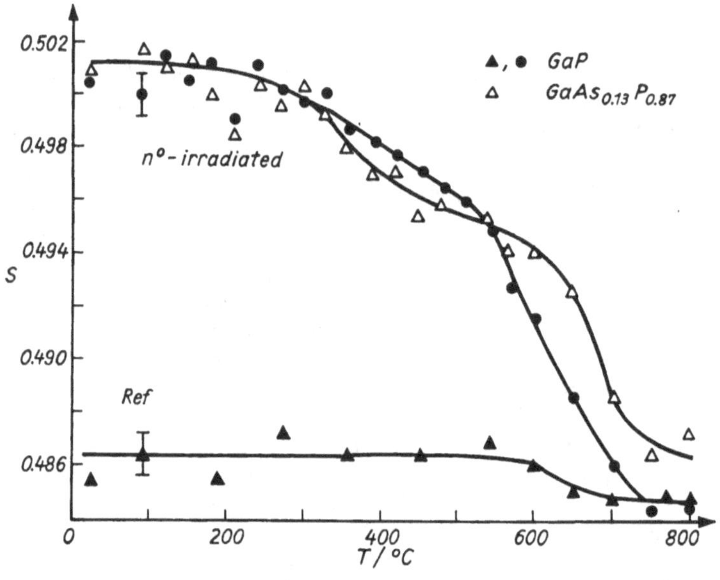

Fig. 2. S parameter of Doppler-broadened annihilation line as function of the isochronal annealing temperature (0.5 hours) in GaP and $GaAs_{0.13}P_{0.87}$ irradiated with 8×10^{18} cm^{-2} fast neutrons

Isochronal annealing of S parameters is shown in Figure 2. There are two annealing stages. In GaP a smooth decrease of S is observed between 200 and 550 °C, followed by a steeper stage at 550–750 °C. In $GaAs_{0.13}P_{0.87}$ the first stage shifts slightly to lower temperatures, while the upper stage extends from 600 to 800 °C. The lifetime decreases from 250 ps to 225 ps after annealing at 800 °C. Using electrical measurements BARNES has found annealing stages at 200, 300, 600, and 800 K in neutron irradiated GaP. The nature of the recovery process occuring in the various stages, however, is not well known. Following the discussion of MARTIN and MAKRAM-EBEID for GaAs the upper stage at 500—800 °C can be attributed to a long-range migration of point defects. In this stage displacement cascades dissolve.

Electron spin resonance (ESR) and deep level transient spectroscopy (DLTS) experiments (MOONEY et al. 1983) have shown that in p-type GaP irradiated with 2 MeV electrons Ga vacancies anneal out at 300—400 °C. In p-type GaP irradiated with 14 MeV electrons recovery stages at 140 °C (250 °C) and 500 °C are seen in the carrier concentration (BRAILOWSKI et al.). Annealing stages in peak height or S parameter at 150—200 °C and 300—400 °C or 500—600 °C are observed in GaP irradiated with 2.3 MeV electrons (BRUDNYI et al.) or 4 MeV protons (DLUBEK et al., to be published), respectively, while in GaP irradiated with 22 MeV alpha particles a slight increase of S between 200 and 400 °C is followed by a sharp decrease above 400 °C (GUPTA et al.). From these results we may conclude that Ga vacancies V_{Ga} disappear in the stage between 200 and 550 °C. The stage at 550—750 °C can be attributed to annealing of Ga vacancies appearing in a more stable configuration e.g. V_{Ga} associated with other point defects inside the displacement cascades. The shift of this stage to higher temperatures in $GaAs_{0.13}P_{0.87}$ may be attributed to a lower long-range mobility of point defects in the disordered alloy. The shift of the lower stage to lower temperatures in $GaAs_{0.13}P_{0.87}$ corresponds to the tendency observed by BRAILOWSKI et al. in electron irradiated material: As-rich $GaAs_{1-x}P_x$ anneals similarly to GaAs and P-rich $GaAs_{1-x}P_x$ anneals similarly to GaP. Neutron-irradiated GaAs exhibits no lower annealing stage, which may indicate that this stage is situated below the irradiation temperature of 90 °C (DLUBEK et al. 1986c).

6. Conclusion

Positron annihilation technique is capable to detect vacancytype defects in GaP and related compounds. In $GaAs_{1-x}P_x$ a linear decrease of $\bar{\tau}$ and S with the composition x of the ternary compound is observed. As-grown GaP may contain vacancies which are interpreted as stoichiometric vacancies in the P sublattice. In neutron irradiated GaP and $GaAs_{0.13}P_{0.87}$ positrons are trapped by negatively charged Ga vacancies. The isochronal annealing curves show two stages at 200—550 °C and 550—800 °C which are attributed to the annealing of Ga vacancies. The characteristic lifetimes of P vacancies and Ga vacancies are found to 290 ps ($\tau_d/\tau_b = 1.30$) and 250 ps ($\tau_d/\tau_b = 1.12$), resp..

References

BARNES, C. E.: Appl. Phys. Lett. **20** (1972) 110

BRAILOWSKY, E. YU., GRIFORGAN, N. E., MARCHOUK, N. P., PAMBUHCHYEN, N. H., TARTACHNIK, V. P.: Ind. Phys. Conf. Ser. 46, (1979) 369

BRUDNYI, V. N., VOROBIEV, S. A., TSOI, A. A., SHAHOVISON, N. I.: Rad. Eff. **79** (1983) 123

DLUBEK, G., BRÜMMER, O., POLITY, A.: Appl. Phys. Lett **49** (1986a) 385

DLUBEK, G., BRÜMMER, O., PLAZAOLA, F., HAUTOJÄRVI, P.: J. Phys. **C19** (1986b) 331

DLUBEK, G., KRAUSE, R., BRÜMMER, O., TITTES, J.: Appl. Phys. **A41** (1986b) 125

GUPTA, A. SEN, NAIDEN, S. V., SEN, P.: Phys. Lett 112A, 399 (1985)

HAUTOJÄRVI, P., MOSER, P., STUCKY, M., CORBEL, C., PLAZAOLA, F.: Apply. Phys. Lett. **48** (1986) 809

JORDAN, A. S., VON NEIDA, A. R., CARNSER, R., KIM, S. K.: J. Electron Soc. **121** (1974) 153

KAUFMANN, U., KENNEDY, T. A.: J. Electron Mater. **10** (1981) 347

MARTIN, C. M., MAKRAM-EBEID, S.: Physica **116** B + C (1983) 371

MOONEY, P. M., KENNEDY, T. A., SMALL, M. B.: Physica **116** B + C, (1983) 431
MOONEY, P. M., KENNEDY, T. A.: J. Phys. **C17** (1984) 6277
SHARMA, S. DAS, MADHUKAR, A.: Phys. Rev. **24** (1981) 2051
VAN VECHTEN, J. A.: J. Electron Mater. **4** (1975) 11

(Received June 16, 1987)

Authors' addresses:
Prof. Dr. sc. G. DLUBEK
Sektion Mathematik/Physik, Pädagogische Hochschule „N. K. Krupskaja"
Halle DDR-4020
Kröllwitzer Str. 44

Prof. Dr. sc. O. BRÜMMER, Dipl.-Phys. A. POLITY
Sektion Physik, Martin-Luther-Universität Halle-Wittenberg
Friedemann-Bach-Platz 6
Halle DDR-4020

Cryst. Res. Technol.	23	1988	2	259–261

J. P. WALCZAK, M. KAMINSKA, J. M. BARANOWSKI

Institute of Experimental Physics,
Warsaw University, Warsaw, Poland

Electrical Properties of GaAs Induced by the Metastability of EL2 Defect[1])

Metastable properties of EL2 defect in undoped semiinsulating GaAs have been studied using photocurrent and thermally stimulated current techniques (TSC). The experimental data suggest that light induced defect reaction takes place when EL2 is transferred to its metastable state.

Introduction

It is widely accepted that EL2, the main deep trap in as-grown GaAs is associated with antistructure arsenic. However, the question if it is an isolated defect or a complex still remains open.

The identification of EL2 as an isolated antisite arsenic defect is strongly supported by the measurements of zerophonon line of EL2 intracenter transition (KAMINSKA et al. 1983) under uniaxial stress (KAMINSKA et al. 1985). Those measurements allowed to prove that EL2 has tetrahedral symmetry. The experimental results lead to basic problem in a theoretical description of EL2, because any model of EL2 has to explain the peculiar behaviour which is observed at low (helium) temperature under illumination with 1 μm light when transfer of that defect from its normal to metastable state occurs. EL2 remains in the metastable state until the sample is heated up to about 40 K for n-type and about 125 K for Sl material (TRAUTMAN et al.).

On the basis of experimental results reported in this paper the idea of light induced defect reaction as responsible for EL2 metastable transition is proposed. The model is founded on the base of chemical reaction of photodissociation and it allows to explain the metastable transition with assumption that EL2 is an isolated antisite arsenic in its ground state.

Experimental and results

Undoped semiinsulating (SI) Liquid Encapsulated Czochralski crystals were investigated. For electrical measurements an ammeter with 10 pA resolution at the finest range was used. Samples were mounted in CF-100 Oxford Instrument Cryostat which made it possible to vary temperature between 10 K and 300 K.

Photocurrent (PC) spectrum of Sl GaAs taken at 300 K is shown in Figure 1.

The shape of PC curve is similar to EL2 absorption spectrum (KAMINSKA et al. 1983). At low temperatures ($T = 10$ K) the time of PC response is very long (approxi-

[1]) Contributed paper presented during the European Meeting on Positron Studies of Defects (PSD 87), March 23–27, 1987, Wernigerode, GDR.

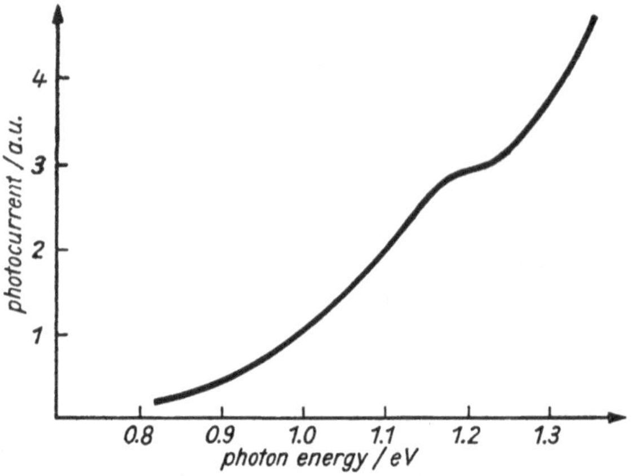

Fig. 1. Photocurrent spectrum measured in undoped SI GaAs crystal at 300 K

mately one minute) which makes the measurements of the whole spectrum difficult However, it is easy to see the quenching of PC spectrum at 10 K after illumination with 1.0 μm light. During a slow heating of the sample, the PC recovery is observed around 125 K after full quenching PC spectra (similarly as in the case of absorption spectrum recovery for SI GaAs), (TRAUTMAN et al.) and around 60 K after partial bleaching of PC spectra (see Fig. 2).

New effect associated with EL2 metastable state has been observed when the thermally stimulated current (TSC) techniques was used. The TSC spectrum was recorded after 1.0 μm illumination of GaAs sample at 10 K. Two different results of TSC measurements have been obtained depending on the time of 1.0 μm illumination. When TSC was observed exactly after the full quenching of PC no TSC up to 130 K has been recorded. On the other hand, when the time of illumination was longer, a peak of TSC at 85 K has been seen (see Fig. 3).

The intensity of TSC peak increasess lowly with the time of illumination and saturates after about 20 min. Photon flux was about 10^{14} cm^{-2} s^{-1}. By means of Hall effect measurements it was found that p-type conductivity is connected with this TSC peak. From TSC spectrum obtained under steady heating rate it is possible to calculate the energy position of the corresponding acceptor level which is 50 $\pm$ 10 meV above the valence band. No TSC peak at 85 K has been recorded when the sample was illuminated with light close to the GaAs band edge or when the energy of photons was below EL2 zero-phonon line energy. On the other hand such light does not produce EL2 metastability in GaAs (VINCENT, BOIS).

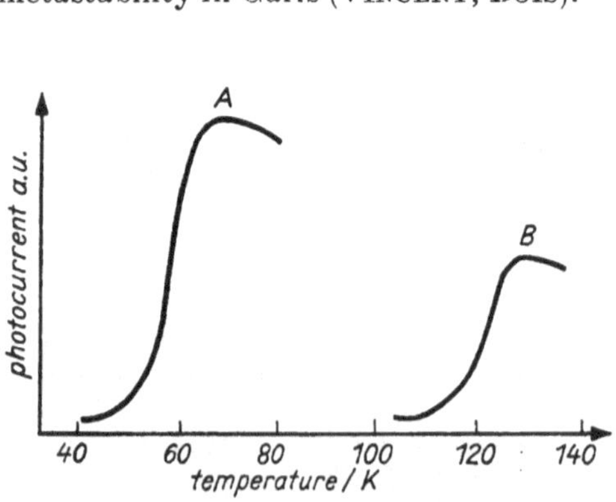

Fig. 2. Recovery of photocurrent (PC) related to EL2 in GaAs measured at 1.0 μm during slow heating of the crystal. Curve A – after partial and curve B – after full quenching PC spectrum

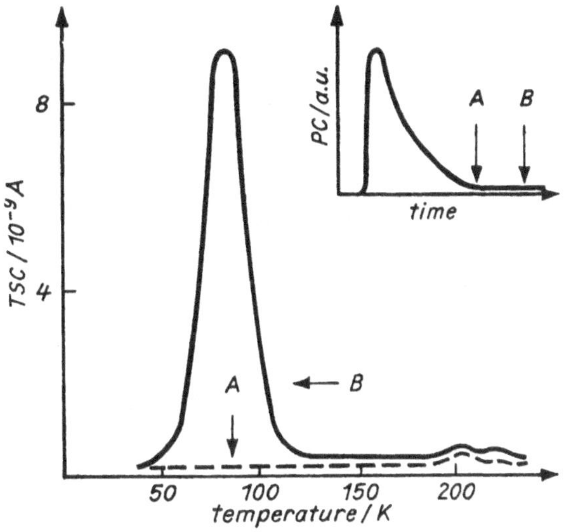

Fig. 3. Thermally stimulated current (TSC) measured after short (A, dotted curve) and long (B, full curve) illumination time. In the insert: photocurrent measured versus time; arrows indicate short (A) and long (B) time of illumination

Discussion

The results of uniaxial stress experiments (KAMINSKA et al. 1985) exclude the possibility of EL2 being the complex with strong binding axial force. The last results of absorption measurements (BARANOWSKI et al.) show no evidence for existence of any weak complexes involving antisite arsenic. In the above experiments EL2 was in its normal state. On the other hand, the new TSC result seems to indicate that light induced defect reaction takes place while EL2 is transferred to the metastable state. After long time illumination with 1 μm light a new acceptor level of energy of about 50 meV above the valence band is created. It must be stressed existence of this new acceptor takes place exclusively after 1 μm illumination, when EL2 is in metastable state.

Therefore we conclude that EL2 is an isolated antisite arsenic. The metastable properties of the defect we explain by defect reaction.

References

BARANOWSKI, J., KAMINSKA, M., KUSZKO, W., WALCZAK, J. P., TRAUTMAN, P., JEZEWSKI, M.: Proc. 14th International Conf. on Defects in Semiconductors, edited by H. J. von BARDELEBEN, Materials Science Forum **10** (1986) 317

KAMINSKA, M., SKOWRONSKI, M., LAGOWSKI, J., PARSEY, J. M., GATOS, H. C.: Appl. Phys. Lett. **43** (1983) 302

KAMINSKA, M., SKOWRONSKI, M., KUSZKO, W.: Phys. Rev. Lett. **55** (1985) 2204

TRAUTMAN, P., KAMINSKA, M., BARANOWSKI, J.: Acta Phys. Polonica **A71** (1987) 269

VINCENT, G., BOIS, D.: Solid State Commun. **27** (1978) 431

(Received June 16, 1987)

Authors' address:
c/o J. P. WALCZAK
Instytut Fizyki Doswiadczalnej, Uniwersytet Warszawski
Hoza 69
00-681 Warszawa, Poland

Cryst. Res. Technol.	23	1988	2	263–267

S. R. DHANALAKSHMI, T. NAGARAJAN, S. RAMASAMY

School of Physics, University of Madras, Madras, India

Dependence of Positron Capture Rate on Dislocation Density in Solid Solutions of KCl–KBr[1])

Generally well-annealed alkali halide crystals exhibit three lifetime components. τ_1 arises due to annihilation with anion and τ_2 due to Bloch Ps. The origin of τ_3 is not well understood. Positron lifetime and etching techniques have been performed on solid solutions of $KCl_{100-x}Br_x$ with mole % of $x = 0.0$, 15, 30, 37, 50, 70, and 100. The positron lifetime spectra have been analysed to resolve into three components. The positron capture rate K_3 for the longest lived component and the dislocation density shows a non-linear variation with composition. The positron capture rate varies linearly with the dislocation density of the samples. The origin of τ_3 is identified as due to positronium formation in the dislocations present in the samples. The trapping probability per dislocation has been estimated to be 2.2×10^{-12}/s.

1. Introduction

The family of alkali halides represents the most typical and intensely studied group of ionic crystals. Positron annihilation technique is very sensitive to the different types of lattice defects present in them. Positron lifetime spectra in alkali halide have been resolved into two or three components by different workers. DANNEFAER et al. have reported three lifetimes.

This longest lifetime is attributed to annihilation in colour centres. This result is consistent with the earlier work of BERTOLACCINI and DUPASQUIER. In the two component spectra the second lifetime is attributed to delocalised Ps state as it undergoes magnetic quenching. We shall not discuss here the longest lived component 1 ns which arises from extrinsic defects like F-centres. In the present analysis as well as in our earlier work (CHINNUSAMY, NAGARAJAN) we have resolved the lifetime component $\tau_2 \sim 500$ ps into two components $\tau_2 \sim 500$ ps and $\tau_3 \sim 800$ ps. In the pioneering work of HYODO, KASAI and FUJIWARA it is well established that τ_2 is due to the delocalised Ps state which on increasing the temperature gradually gets into a trapped state. They have shown that the τ_2 component in their analysis (τ_2 and τ_3 are not resolved) exhibit magnetic quenching effect and this is ascribed to delocalised Ps state. We have measured positron lifetime in well annealed crystals of KCl and KBr with a good lifetime set up with 240 ps and have observed three lifetime components $\tau_1 \sim 160$ ps with intensity $I_1 \sim 30\%$, $\tau_2 \sim 500$ ps with intensity $I_2 \sim 50\%$ and $\tau_3 \sim 800$ ps with $I_3 \sim 20\%$. The results agree very well with the results of CHINNUSAMY et al. on pure and doped KCl single crystals. None of the earlier workers have carried out a microstructure analysis in these crystals.

The study of alkali halide solid solutions is interesting in view of the nature of defects present in them and their role in understanding the various physical properties

[1]) Contributed paper presented during the European Meeting on Positron Studies of Defects (PSD 87), March 23–27, 1987, Wernigerode, GDR.

associated with them. Physical properties like macrodensity, intensity of coloration (Smakula et al.), dielectric constant (Kamiyoshi, Nigara), electrical conductivity, microhardness, radiation hardening (Haribabu et al.; Subba Rao, Haribabu 1978a, b; Veeresham et al.) vary non-linearly with composition, whereas the diamagnetic susceptibility (Van Den Bosch et al.) and lattice parameters closely follow a linear variation with composition (Haribabu et al.). Conduction of ionic crystals is a defect-controlled property. The defect concentration increases exponentially with rise of temperature and the electrical conductivity increases correspondingly. The ionic conductivity measurements on solid solutions of KCl–KBr by Schultze indicated that the concentration of vacancies in mixed crystals slightly exceed the pure components. Smakula et al. investigated macrodensity, microhardness and intensity of coloration in KCl–KBr mixed crystals. The macrodensity determined by weighing showed marked deviation from linearity. These results indicated that these crystals may contain either vacancies or interstitials. These two types of defects are differentiated by determining colour centre density measurements. The increase of F-band with radiation dose for KCl was lowest, for KBr highest and for mixed crystals was in between. These results indicated that the number of vacancies in mixed crystals did not increase substantially. The concentration of vacancies was found to be of the same order of magnitude in the mixed crystals as the end components.

The studies on F-Centre formation in alkali halide mixed crystals by Arends et al. indicated that the concentration of vacancies in mixed crystals exceed that in pure end components only slightly. Zavodovskaya et al. investigated the electrical conductivity of electron bombarded mono-crystals of solid solutions of KCl–RbCl. Temperature variation of electrical conductivity of equimolar solid solutions of non-irradiated KCl–RbCl and RbCl crystals with Sr^{+} impurity has been found to be less than specimens irradiated in various doses at temperature close to room temperature. From the investigations they concluded that the temperature variations of electrical conductivity of mixed crystals was more in magnitude than the pure components which has been irradiated in a similar way. Further KCl–RbCl system was found to be less defective in nature compared to KCl–KBr system indicating that the change in anion produces more effect. Haribabu, Subba Rao (1978b) et al. have studied the conductivity in mixed crystals of KCl and KBr and KBr–KI systems.

The results indicated non-linear variation of ionic conductivity and activation energy and also the conductivity anisotropy along parallel and perpendicular to growth axis has been explained as due to the enhanced diffusion of charge carriers along dislocations and grain boundaries which are more in mixed crystals.

We have investigated the dislocation density and its distribution in mixed crystals as a function of composition in $KCl_{100-x}Br_x$, $x = 0$, 15, 30, 37, 50, 70, and 100 using etching technique. The dislocation density was found to be more in mixed crystals compared to their end products which is consistent with result previously reported by Veersham et al. for KCl–KBr system. From positron lifetime measurements the capture rate for the longest lifetime compound has been calculated. This showed a non-linear variation with composition similar to that of the dislocation density which resulted in a correlation between dislocation density and capture rate K_3.

2. Experimental

Positron lifetime measurements have been made with the conventional fast slow coincidence system. The prompt curve of the apparatus with a Co^{60} gamma source and a Na^{-22} energy window setting had a FWHM of 240 ps. The positron source was made of a neutral carrier-free solution deposited in a nickel foil (1.1 mg/cm²). The samples were single crys-

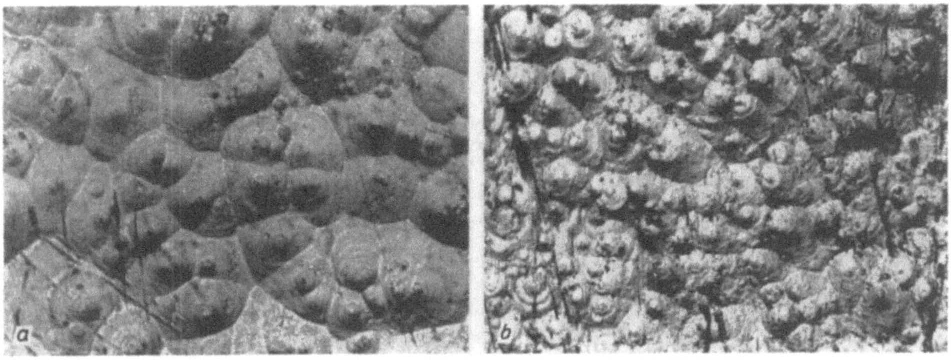

Fig. 1a) Photomicrograph of $KCl_{70}Br_{30}$, b) Photomicrograph of $KCl_{50}Br_{50}$

stals and the source was sandwitched between two crystal pieces. The measuring time for one lifetime spectrum was 26 hours and approximately 2×10^6 counts were accumulated. The lifetime spectra were analysed with three lifetime components after source and background corrections as a sum of exponential decay components using POSFIT program (ZAVADOVSKAYA et al.).

The density and distribution of dislocations were studied by etch pit technique. The etchant consisted of freshly distilled methanol saturated with lead chloride. The photomicrographs of the etched samples are shown in Figure 1a, b. The capture rate of the third lifetime $K_3 = I_3(\tau_1^{-1} - \tau_3^{-1})$ has been calculated for all the samples, here τ_1, and τ_3 are the short and longest lived lifetime components and I_3 is the intensity of the longest lived component.

3. Results and discussion

The results show that the positron capture rate varies from 0.73 ns^{-1} to 1.36 ns^{-1}. The capture rates vs dislocation density for all the samples has been shown in Figure 2. The capture rate K_3 is found to be a maximum for $KCl_{50}Br_{50}$. The dislocation density for this composition is also found to be a maximum, shown in Figure 1b. The magnitude of the longest lifetime obtained in the present studies ranges from 758 ps to 902 ps. The magnitude reveals that this lifetime may be due to delocalised positronium in all

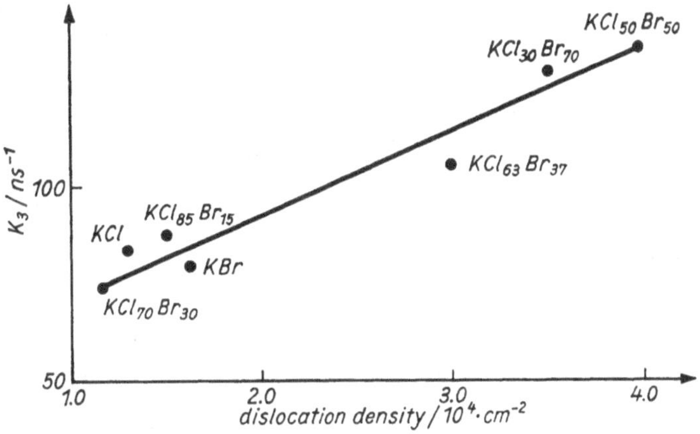

Fig. 2. Linear variation of capture rate K_3 with dislocation density

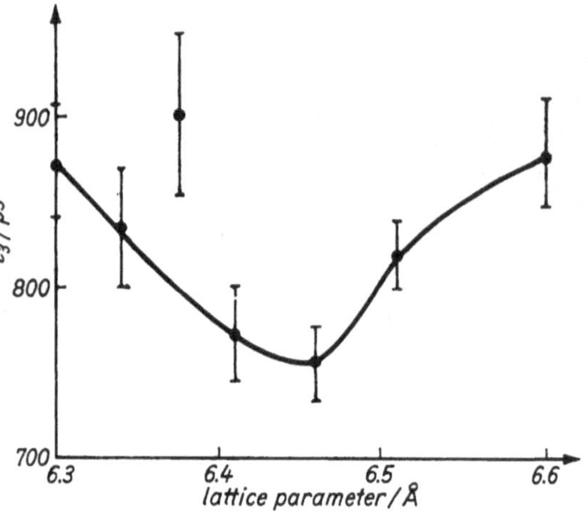

Fig. 3. Variation of τ_3 with lattice parameter

the samples. The origin of the positronium formation is verified by etching technique and the dislocation density has been found to have a linear relation to the positron capture rate. The magnitude of third lifetime τ_3 obtained in the present measurements vary smoothly with the lattice parameter, shown in Figure 3. This smooth variation ensures that dislocations are directly responsible for delocalised positronium formation. The dislocation density estimated for all the samples show a maximum for the composition $KCl_{50}Br_{50}$.

4. Conclusion

The results of the present investigations reveal that all the seven samples have dislocations and their density of distribution varies from a minimum of $1.17 \times 10^4/cm^2$ for the composition $KCl_{70}Br_{30}$ to a maximum of $3.98 \times 10^4/cm^2$ for the composition $KCl_{50}Br_{50}$. The three lifetime components obtained from the present measurements $\tau_1 \sim 160$ ps, $\tau_2 \sim 500$ ps and $\tau_3 \sim 800$ ps with intensities $I \sim 30\%$, $I_2 \sim 50\%$ and $I_3 \sim 20\%$. The short lived components τ_1 and τ_2 do not seem to vary much with composition, whereas the longest lived component τ_3 has a systematic trend with composition. The linear variation of the capture rate K_3 of the third lifetime component τ_3 with dislocation density indicates that the positronium formation is in the dislocation present in these samples. The smooth variation of τ_3 with lattice parameter again ensures that these dislocations are directly responsible for positronium formation.

The authors wish to express their sincere gratitude to Prof. Y. V. G. S. Murty, Department of Physics, I. I. T. Madras and Prof. P. Ramasamy, Crystal Growth Centre, Anna University, Madras for their helpful collaboration: Work supported by UGC COSIST INDIA.

References

ARENDS, J., DEN HARTOG, H. W., DEKKAR, A. J.: phys. stat. solidi **10** (1965) 105
BERTOLACCINI, M., DUPASQUIER, A.: Phys. Rev. **B1** (1970) 2896
CHINNUSAMY, M., NAGARAJAN, T.: phys. stat. solidi (b) **112** (1982) 297
DANNEFAER, S., DEAN, G. W., HOGG, B. G.: Phys. Rev. **B13** (1976) 3715
FUJIWARA, K., HYODO, T., TAKAKUSA, Y., SUEOKA, O.: J. Phys. Soc. Japn. **39** (1975) 403

HARIBABU, V, SUBBARAO, U. V., VENKARA RAMIAH, K.: phys. stat. solidi **28** (1975) 269
HYODO, T., TAKAKUSA, Y.: J. Phys. Soc. Jpn. **42** (1977) 1065
HYODO, T., TAKAKUSA, Y.: J. Phys. Soc. Jpn. **45** (1978) 795
KAMIYOSHI, K., NIGARA, Y.: phys. stat. solidi **6** (1971) 223
KASAI, J., FUJIWARA, K.: J. Phys. Soc. Jpn. **51** (1982) 3077
KASAI, J., HYODO, T., FUJIWARA, K.: J. Phys. Soc. Jpn. **52** (1983) 3671
KIRKEGAARD, P., ELDRUP, M.: Comput. Phys. Commun. **7** (1974) 401
SCHULTZ, H.: Thesis, University of Göttingen 1952
SMAKULA, A., MAYNARD, N. C., REPUCCI, A.: Phys. Rev. **130** (1963) 113
SUBBA RAO, U. V., HARIBABU, V.: Pramana **11** (1978) 149
SUBBA RAO, U. V., HARIBABU, V.: Crystal lattice Defects **8** (1978) 21b
VAN DEN BOSCH, A., DRESSELAERS, J., VANSUMMEREN, J., HOVI, M.: phys. stat. solidi **11** (1972) 479
VEERESHAM, P., SUBBA RAO, U. V., HARIBABU, V.: J. Cryst. Growth **52** (1981) 943
ZAVADOVSKAYA, E. K., IVANKIUNA, J. S., IYA MELIK-GAIKAZYAN: IZV VYUS Ucheb-Zab. Fiz. **12** (1970) 92

(Received June 16, 1987)

Authors' address:
S. R. DHANALAKSHMI, T. NAGARAJAN, S. RAMASAMY
School of Physics, University of Madras
Madras 600025, India

Cryst. Res. Technol. | 23 | 1988 | 2 | 269–272

UDO MARX

Martin-Luther-Universität Halle-Wittenberg, Sektion Physik, Halle, GDR

Positron Studies of $BaTiO_3$[1])

Measurements of positron lifetime, Doppler broadening and angular correlation have been taken from n-type sintered ceramics on the basis of $BaTiO_3$. Polycrystalline n-doped $BaTiO_3$ shows a pronounced positive temperature coefficient (PTC) of electrical resistance, e.g. an increase of resistivity by several orders of magnitude within a temperature interval of a few degrees in the vicinity of the ferroelectric phase transition at 130 °C centigrade (eg. HEYWANG). This little-understood effect usually is discussed in terms of potential barriers resulting from an inhomogeneous distribution of Ba vacancies near grain boundaries (a review of models is given by LEWIS et al.).

The present investigations performed by positron annihilation at technical PTC ceramics in as-received state and after special ageing cycles give evidence of existence of vacancy-type defects becoming mobile above 700 ... 800° centigrade. High-temperature ageing as well as long sintering intervals obviously lead to a homogeneous distribution of the defects that are thought to be Ba vacancies.

Neither lifetime spectra nor angular correlation measurements with magnetic field gave an indication of positronium formation in the materials studied. Just as little the ferroelectric phase transition came to light in the parameters of angular correlation curves measured below and above the Curie point.

Experimental results

Positron lifetime measurements have been taken from technical sintered ceramic samples of $(BaCaPb)TiO_3$ with varying La and Nb dopant concentration at room temperature in as-received state (Fig. 1). Clearly, two lifetime components can be resolved, $\tau_1 = 160 ... 165$ ps, $\tau_2 \approx 280$ ps.

The intensity of the τ_2 component increases with Nb concentration whereas in the case of La doped material there seems to be only a very slight variation of I_2. In the perovskite lattice La and Nb are known to occupy A (Ba) sites and B (Ti) sites, respectively.

Two series of $(BaCaPb)TiO_3$ ceramics each with different La- and Nb-doping have been successively aged at 800, 1000, and 1200 °C for two hours whereafter they have been cooled down either slowly (30 ... 50 K/h), or rapidly ($\approx$ 1000 K/min). Results of positron lifetime measurements after each step are shown in Figure 2. Results that have been received after ageing below 800 °C generally coincide with those from original state and, therefore, are not indicated in the figure. An influence of ageing begins to manifest at 800 °C, the general trend being an increase of I_2.

---

[1]) Contributed paper presented during the European Meeting on Positron Studies of Defects (PSD 87), March 23–27, 1987, Wernigerode, GDR.

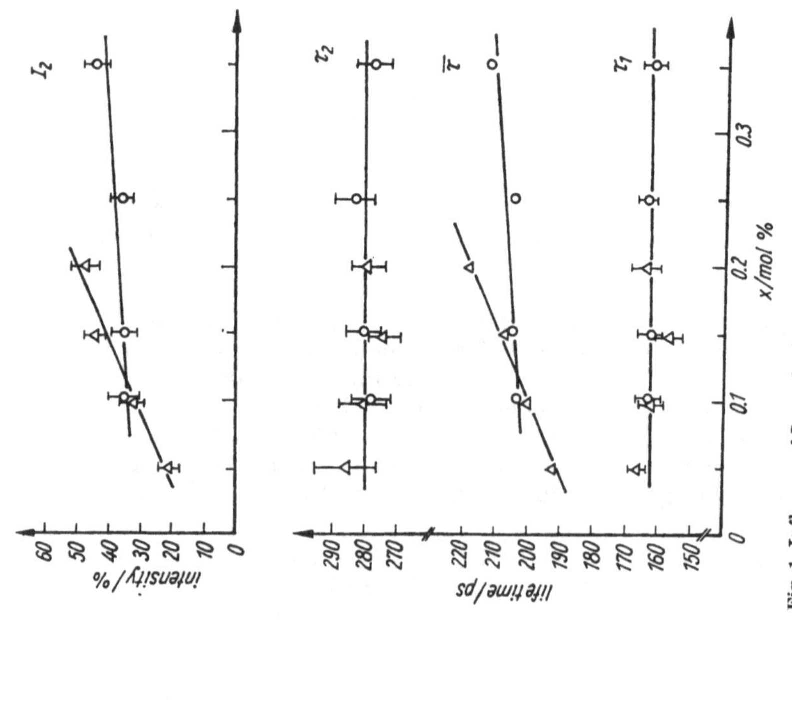

Fig. 1. Influence of La and Nb dopant concentration in technical sintered ceramics of (BaCaPb) TiO_3 + **x**M-type on positron lifetime parameters τ_1, τ_2, $\bar{\tau}$, and I_2 at room temperature. ○ M = La, △ M = Nb

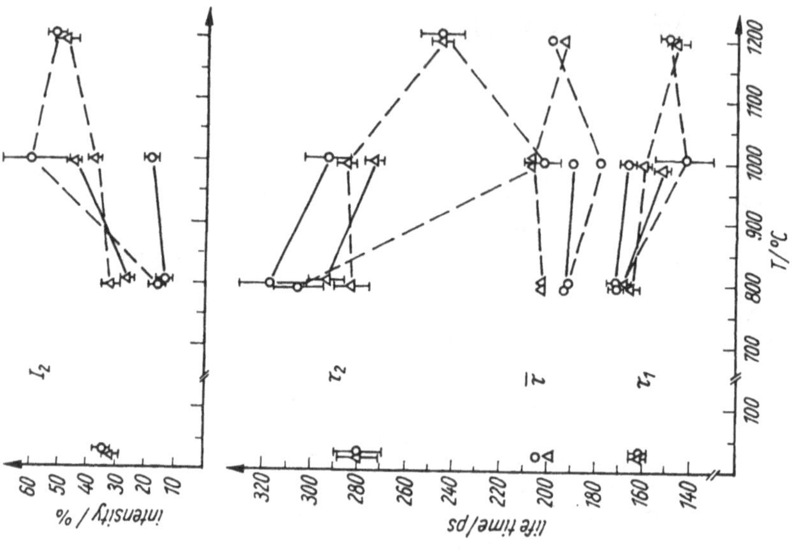

Fig. 2. Positron lifetime parameters τ_1, τ_2, $\bar{\tau}$, and I_2 vs ageing temperature. ○ (BaCaPb)TiO_3 + 0.15 % La, △ (BaCaPb)TiO_3 + 0.10 % Nb. ——— after slow cooling, – – – after quenching

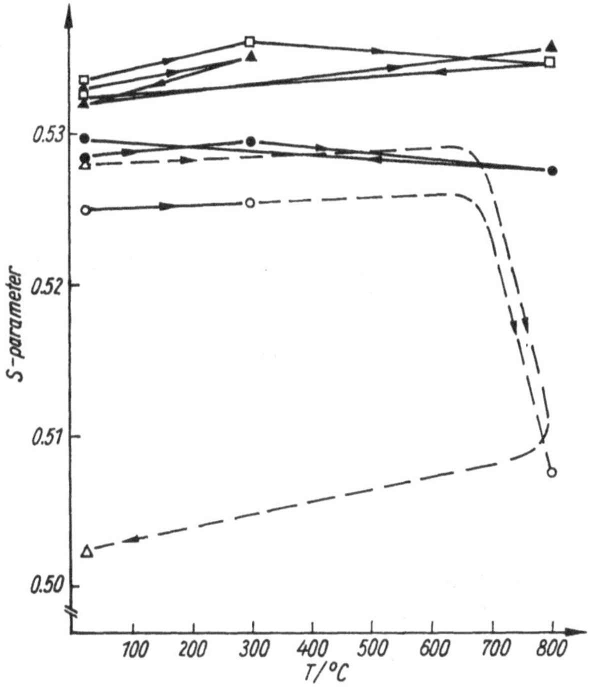

Fig. 3. Variation of S-parameter from angular correlation measurements during temperature cycles 25° → 300° (→ 25° for ▲) → 800° → 25° C for (BaCaPb)TiO_3 + xM. ○ xM = 0.15 % La, △ xM = 0.10 % Nb, both after pre-ageing at 1200 °C for 2 hours, ● xM = 0.35 % La, ▲ xM = = 0.20 % Nb, □ ZnO varistor ceramics

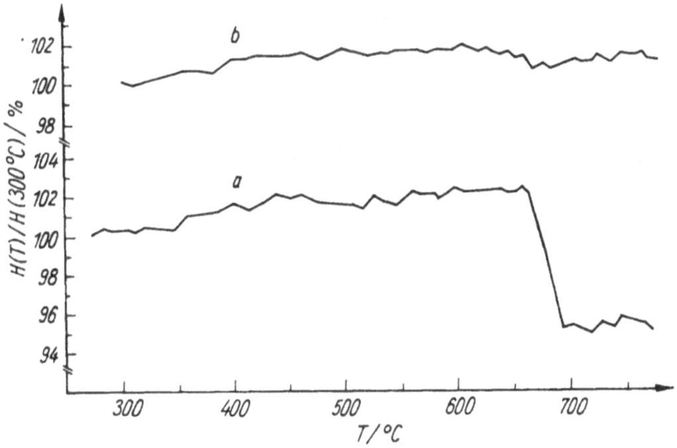

Fig. 4. Angular correlation peak height vs temperature for (BaCaPb)TiO_3 + 0.15 % La. a) after pre-ageing at 1200 °C for 2 hours and b) in as-received technically sintered state

S-parameter and peak height vs temperature determinations have been performed from angular correlation experiments. All samples that had undergone the temperature cycle starting from original as-received state only showed the expected reversible temperature effect due to lattice expansion.

Those samples that had been pre-aged at 1200 °C for 2 hours, though, behaved quite differently as sketched out in Figure 3 and 4.

Conclusions

The results can be understood in the light of a model based on an inhomogeneous distribution of defects (DANIELS et al.). The sintering and cooling processes in technical PTC ceramics maufacturing are controlled in such manner that defect migration away from grain boundaries will stop after a defective region in the order of 1 μm thickness has formed. Therefore, the effective volume containing a detectable concentration of this kind of defects that usually are though to be Ba vacancies is relatively small in the as-received state of the samples. At about 800 °C these defects become mobile and start to spread over the whole crystallite volume thus flattening the concentration profile and destroying the PTC effect.

The positron annihilation results confirm this concept insofar the long-lived component, τ_2, can be attributed to annihilation in defects of cation vacancy type. Changes in annihilation parameters then should primarily be due to vacancy migration at sufficiently high temperature. Furthermore, defect reactions like formation of complexes have to be taken into consideration. To get beyond this point in interpretation more insight into basic processes underlying positron annihilation in semiconducting oxide ceramics must be gained. First of all, the positron diffusion length is subject to further investigations that call for well-characterized laboratory made ceramic and single crystalline samples of simple constitution.

References

DANIELS, J., HÄRDTL, K. H., WERNICKE, R.: Philips techn. Rundsch. **38** (1979) 1
HEYWANG, W.: J. Mat. Sci. **6** (1971) 1214
LEWIS, G. V. et al.: J. Amer. Ceram. Soc. **68** (1985) 555

(Received August 10, 1987)

Author's address:
Udo MARX
Sektion Physik der Martin-Luther-Universität Halle-Wittenberg
Friedemann-Bach-Platz 6
Halle
DDR-4010

Cryst. Res. Technol.	23	1988	2	273–277

G. BRAUER, G. BODEN, T. HÜBERT, M. KIRSCH

Academy of Sciences of the GDR, Central Institute for Nuclear Research
Rossendorf, Dresden, and
Academy of Sciences of the GDR, Centra l Institute for Inorganic Chemistry, Berlin-DDR

Sintering and Crystallization of Titanium-containing Lead Borate Glasses[1])

The sintering and crystallization of glass powders with chemical composition 47 PbO–38 B_2O_3–15 TiO_2 (mole%) was investigated under non-isothermal conditions by shrinkage measurements, differential thermo-analysis (DTA), X-ray diffraction (XRD), dielectric measurement and positron annihilation (Doppler broadening). The correlation between changes of macroscopic properties and positron annihilation results recommends this method for the detection of structural changes in glasses.

Sinterung und Kristallisation von Glaspulvern der chemischen Zusammensetzung 47 PbO–38 B_2O_3–15 TiO_2 (Mol%) wurden unter nichtisothermen Bedingungen mittels Schwindungsmessungen, differentieller Thermoanalyse (DTA), Röntgenbeugung (XRD), dielektrischen Messungen sowie Positronenannihilation (Dopplermessungen) untersucht. Die Korrelation zwischen Änderungen makroskopischer Eigenschaften und Parametern der Positronenannihilation empfiehlt diese Methode zum Nachweis struktureller Änderungen in Gläsern.

1. Introduction

Positron annihilation is now a well established tool to study the structural properties of solids (BRANDT, DUPASQUIER). The main aim of investigations performed on inorganic glasses of different kind and composition was to follow phase-separation and crystallization. By BRAUER and BODEN these studies were shortly reviewed and published together with their results obtained on silica glasses.

Glasses and glass-ceramics of the system PbO–TiO_2–Al_2O_3 are of technical interest especially for low melting solder glasses and in thick film hybrid techniques. We followed structural changes during sintering and crystallization of titanium containing lead borate glasses by positron annihilation (Doppler broadening), differential thermo-analysis (DTA), X-ray diffraction (XRD) and dielectric measurements.

2. Experimental

Glass of the composition 47 PbO–38 B_2O_3–15 TiO_2 (mole%) was melted at 1370 K in a platinum crucible. Afterwards it was crusted in an achat mill and classified in order to obtain glass powder fractions of a medium radius $\bar{R}$ in the range (10–200) μm. These fractions were compacted to form pellets (20 mm diameter, 3 mm thickness) before heat treatment in air.

---

[1]) Contributed paper presented during the European Meeting on Positron Studies of Defects (PSD 87), March 23–27, 1987, Wernigerode, GDR.

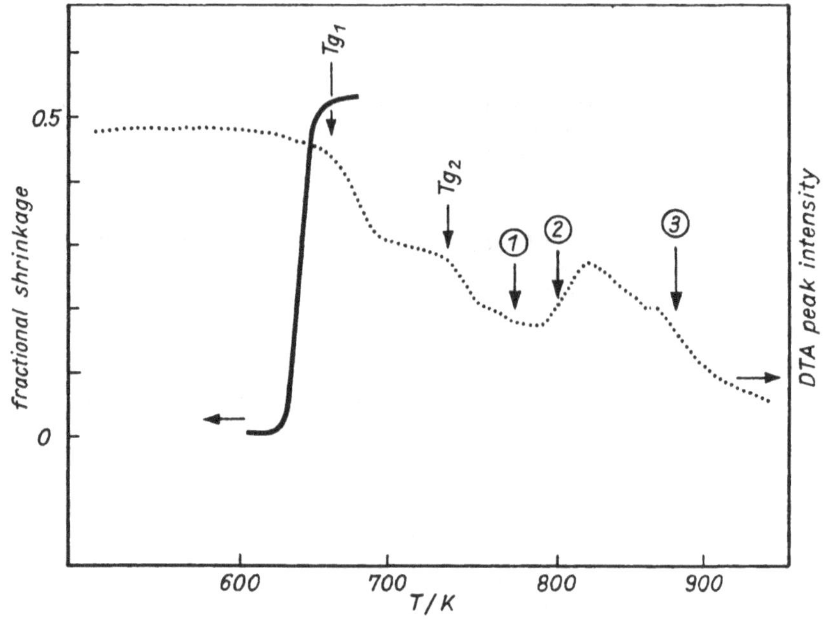

Fig. 1. Fractional shrinkage and DTA peak intensity of a glass powder fraction with $\bar{R} = 54$ μm (heating rate 10 K · min⁻¹)

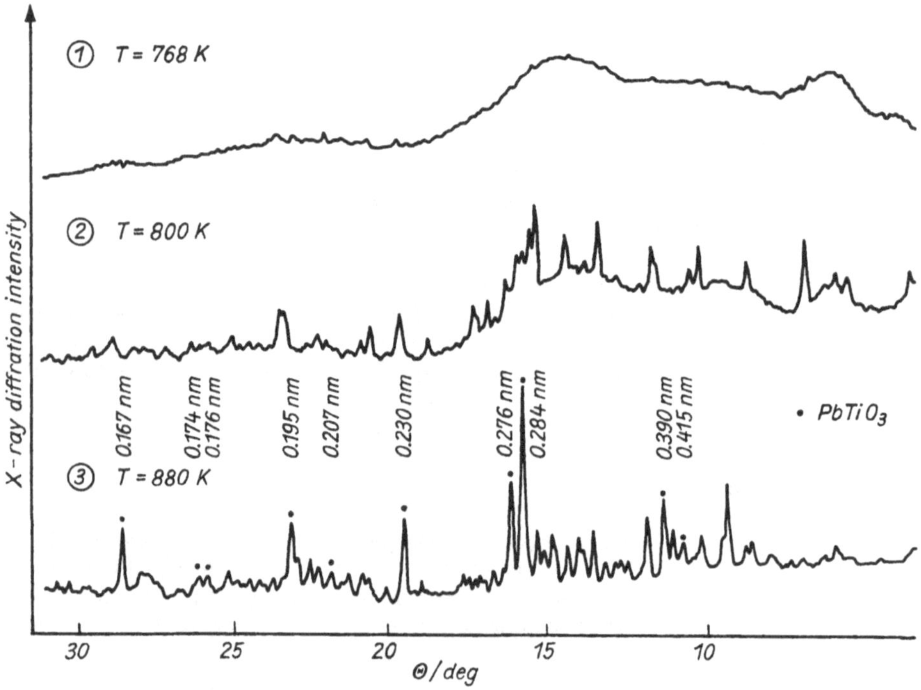

Fig. 2. X-ray diffraction intensity as a function of Bragg angle θ for three different temperatures T reached (heating rate 10 K · min⁻¹) before cooling

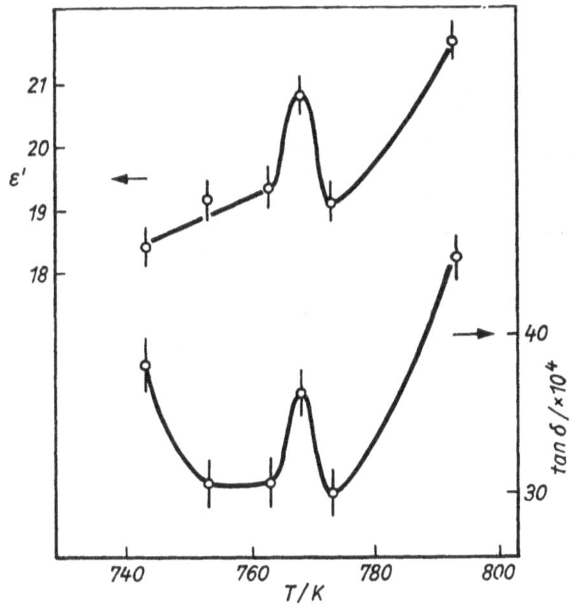

Fig. 3. Real part ε' of the dielectric constant $\varepsilon = \varepsilon' + i\varepsilon''$ and dielectric losses $\tan \delta = \varepsilon'/\varepsilon''$ at 1 MHz as a function of temperature T reached (heating rate $10\ K \cdot min^{-1}$) before cooling. The curves shown are averaged from all powder fractions

Sintering and crystallization generally were carried out under non-isothermal conditions. Samples were always heated with a rate of $10\ K \cdot min^{-1}$ up to a certain temperature $T \leqq 900$ K and immediately quenched after reaching this temperature.

Shrinkage measurements were used to follow the sintering of glass powders. DTA diagrams of glass powders were obtained by use of a NETZSCH 404 M Thermoanalyser.

XRD studies were carried out by use of a Guinier-camera in order to identify crystalline phases.

Dielectric measurements were performed by use of a TESLA RLCG-bridga BM 539 in order to estimate the real part ε' of the dielectric constant ($\varepsilon = \varepsilon' + i\varepsilon''$) as well as the dielectric losses ($\tan \delta = \varepsilon'/\varepsilon''$) at 1 MHz.

The measuring technique of positron annihilation and data treatment has been described in detail elsewhere (Brauer, Boden).

3. Results and discussion

The results are shown in Figure 1—4 and may be summarized as follows:

— Sintering of glass powders starts at about 630 K ($\bar{R} = 54\ \mu m$) and results in a dense amorphous body (Fig. 1). Crystalline parts could not be detected (Fig. 2, curve 1).

— DTA curves exhibit two denothermic peaks (Fig. 1, Tg 1, Tg 2) which indicate phase-separation in the glass (Polyakova, Morosova). Two additional exothermic peaks are caused by crystallization of a not further identified borate phase (Fig. 2, curve 2) and of lead titanate ($PbTiO_3$) (Fig. 2, curve 3).

— Positron annihilation studies (Fig. 4) reveal a relative minimum of the Doppler broadening parameter S at 768 K.

From the shift of the first exothermic peak due to variation of heating rate or grain size it was possible to determine the activation energy of primary crystallization to be about $300\ kJ \cdot mol^{-1}$ (Kissinger; Müller). This value agrees with results of

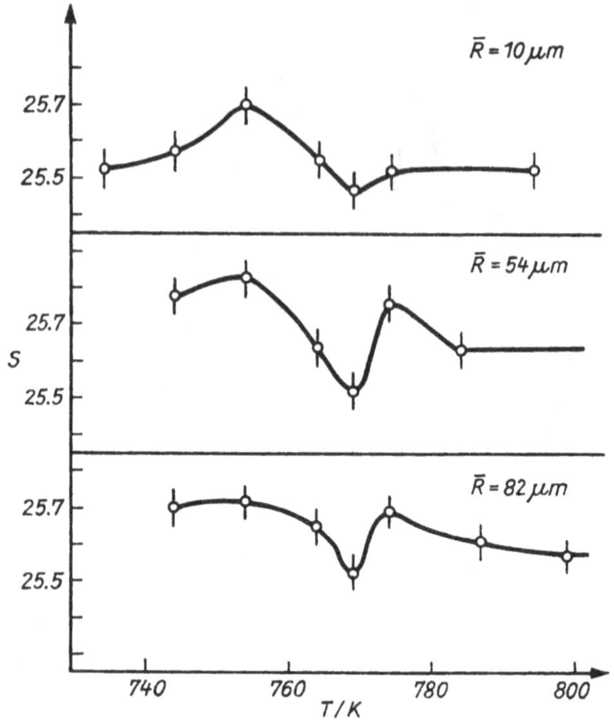

Fig. 4. Doppler broadening parameter S of positron annihilation as a function of temperature T reached (heating rate 10 K · min^{-1}) before cooling for three different powder fractions

Bergeron and Russel obtained for a glass with similar composition. The second exothermic peak at 860 K does not change its position due to variation of heating rate or grain size thus indicating a volume crystallization process of $PbTiO_3$.

The temperature 768 K where the relative maxima of ε', $\tan\delta$, and the relative minima of S are located (Figs. 3, 4) coincides with the Curie temperature of pure $PbTiO_3$, i.e. the change from ferro-electric to para-electric behaviour. Ferro-electric phase transformation was also observed in glassy $LiNbO_3$, $LiTaO_3$, $Pb(Zr, Ti)O_3$, and in the system $PbO–BaO–TiO_2–B_2O_3–SiO_2$ (Kumar et al.). Regarding the interpretation of our results we assume that already in the glass — as probably due to phase separation — regions are built in which undergo a phase transformation like $PbTiO_3$ at 768 K. The instable, unarranged transformation state of these regions must be frozen in by fast cooling of the samples after heating. In this way the results give indications on "pre-ordered states" in titanium-lead-borate glasses.

References

Bergeron, C. G., Russel, C.: J. Amer. Ceram. Soc. **48** (1965) 115

Brandt, W., Dupasquier (eds): Positron Solid-State Physics. Proc. Int. School Physics "Enrico Fermi", Course LXXXIII, Varenna on Lake Como, Italy 1981, ed. Amsterdam 1983

Brauer, G., Boden, G.: ZfK-442 (1981)

Kissinger, H. E.: J. Res. Nat. Bur. Standards **57** (1956) 217

KUMAR, D., PARKASH, O., PANDEY, L., RAJGOPALAN, R.: in: Coll. Papers Fourteenth Int. Congr. on Glass. New Dehli, India 1986, p. 312
MÜLLER, R.: Thesis, Academy of Sciences of the GDR, 1987
POLJAKOVA, I. G. MOROSOVA, E. V.: 8. Vsesoyuznoe soveshchanie po stekloobraznomu sostoyaniyu, Leningrad 1986

(Received June 16, 1987)

Authors' addresses:

Dr. sc. nat. G. BRAUER, Dr. sc. nat. G. BODEN
Academy of Sciences of the GDR
Central Institute for Nuclear Research Rossendorf
PF 19
DDR-8051 Dresden

Dr. rer. nat. T. HÜBERT, Dr.-Ing. M. KIRSCH
Academy of Sciences of the GDR
Central Institute for Inorganic Chemistry
Rudower Chaussee 5
DDR-1199 Berlin

Cryst. Res. Technol.	23	1988	2	279–283

William W. WALKER

Department of Physics & Astronomy, The University of Alabama Tuscaloosa, Alabama, U.S.A.

Positron Lifetime Studies of X-Irradiated N-Acetyl-D,L-Alanine[1]

Polycrystalline samples of N-acetyl-D,L-alanine (AA) which had received an X-ray dose of approximately 6 Mrad have been studied by the positron annihilation technique. Data were obtained as a function of time following the radiation treatment, and also as a function of subsequent annealing temperature. The large decrease of the intensity of the longest lifetime component (I_3) was observed to be reversible when assisted by heat treatment.

Polykristalline Proben von mit ungefähr 6 Mrad Röntgenstrahlen bestrahltem N-acetyl-D,L-alanin (AA) wurden mit der Positronenannihilation-Methode untersucht. Diese Methode ergab Daten als Funktion der Zeit nach der Bestrahlung und auch als Funktion der Temperatur nach späterem Erwärmen. Dann konnte man beobachten, daß die große Intensitätsabnahme der am langsamsten abklingenden Komponente (I_3) bei Erwärmung reversibel ist.

1. Introduction

It has previously been demonstrated (HADLEY, HSU; ELDRUP et al.; JEAN, ACHE; ITO, TABATA) that X-ray or gamma-ray irradiation of organic materials can have a significant effect on the intensity of the longest positron lifetime component in those substances. However, the origin of this effect is not entirely clear. It is well known that X- or gamma-irradiation of organics not only produces free radicals but also structural changes such as swelling and vacancy formation. Since positron annihilation is sensitive to all of these effects, changes in the positron lifetime parameters will be governed by whichever radiation-produced effect is dominant.

Recently KURODA et al. have observed a new effect: the anisotropic expansion of an organic crystal, N-acetyl-D,L-alanine (AA) following irradiation by X-rays. This expansion, which occurs along the c-axis, was found to be as large as 20% when assisted by heat treatment. Since positronium formation and annihilation characteristics can be influenced by changes in the crystal lattice of the medium surrounding a positron, the present work was undertaken in an attempt to observe the crystal expansion of irradiated AA by this technique.

2. Experimental

The polycrystalline AA was obtained from Sigma Chemical Company. Samples were irradiated at room temperature by an X-ray source operating at 45 kV and 30 mA, producing a dose of approximately 3 Mrad per hour in the sample. Positron lifetime measurements were made with a conventional apparatus (WALKER, MUELLER) and the data were analyzed into three components using Positronfit Extended (KIRKEGAARD, ELDRUP).

[1]) Contributed paper presented during the European Meeting on Positron Studies of Defects (PSD 87), March 23–27, 1987, Wernigerode, GDR.

3. Results

In order to determine the irradiation time needed for maximum effect, lifetimes were measured for samples which had been irradiated for varying lengths of time. These results are given in Figure 1 which shows the effect is saturated for irradiation times exceeding about 30 minutes.

One sample of AA was irradiated for 2 hours at room temperature and maintained at that temperature throughout the measurements. Immediately following irradiation, the longest lifetime component (τ_3) showed an increase of about 10% (from 1.16 to 1.28 ns) and the intensity of this component (I_3) showed approximately a 50% decrease (from 31.0% to 16.4%). As this sample has aged, τ_3 has shown little change but I_3 has exhibited gradual changes tending toward the value obtained prior to irradiation (Fig. 2). Although electron spin resonance (ESR) measurements were not made as a function of time, a measurement taken 625 days after irradiation revealed no ESR-detectable radicals.

A second sample of AA was irradiated for 2 hours at room temperature and then placed in an oven for periods of 1 hour at temperatures ranging up to 124 °C. It is

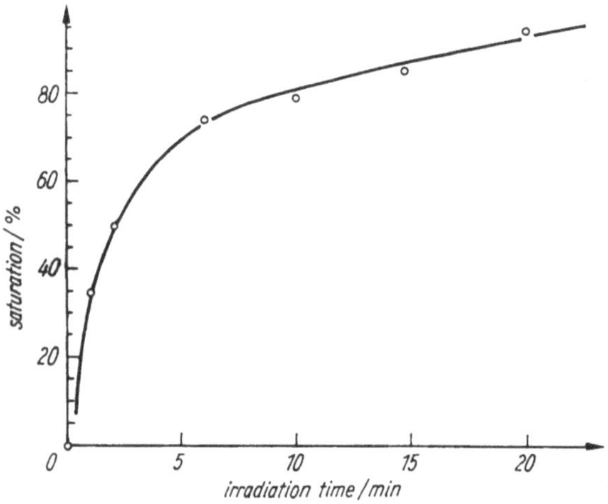

Fig. 1. Comparison of the change in I_3 at short irradiation times with that produced by long-time irradiation, i.e., the saturated value

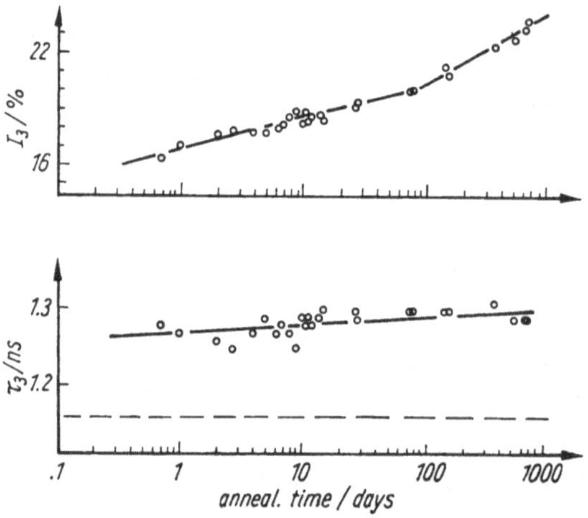

Fig. 2. Time dependence of τ_3 and I_3 following X-irradiation of AA for 2 hours. Sample was maintained at room temperature. Dashed line represents pre-irradiation value

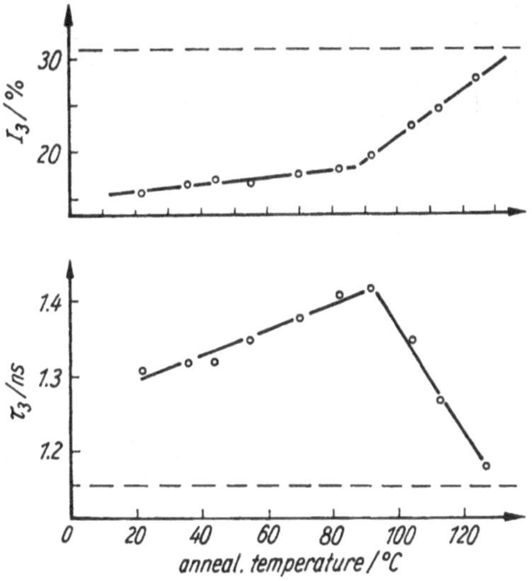

Fig. 3. Dependence of τ_3 and I_3 on annealing temperature for AA sample which had been X-irradiated for 2 hours. Dashed lines represent pre-irradiation values

noted in Figure 3 that both τ_3 and I_3 returned almost to their pre-irradiation values after the sample was annealed at 124 °C.

It is known that following resolidification from the melt, an organic solid may develop lattice defects. Thus, an unirradiated sample of AA was quick-cooled from the melt. The observed changes in the positron lifetime parameters were in the same directions as those observed for irradiated samples, but the magnitudes of these changes were different. τ_3 increased from 1.16 to 1.50 ns while I_3 decreased from 31.0% to 24.0%. Figure 4 shows that, as a function of time elapsed since resolidification, the lifetime parameters gradually changed, tending toward the pre-melt values.

The data of Figure 5 show that heat treatment of a previously melted sample can approximately restore the positron lifetime and intensity to the pre-melt values. Although the annealing temperature required is essentially the same as that observed for the irradiated (but not melted) sample, there are significant differences in the manner in which τ_3 and I_3 vary with annealing temperature.

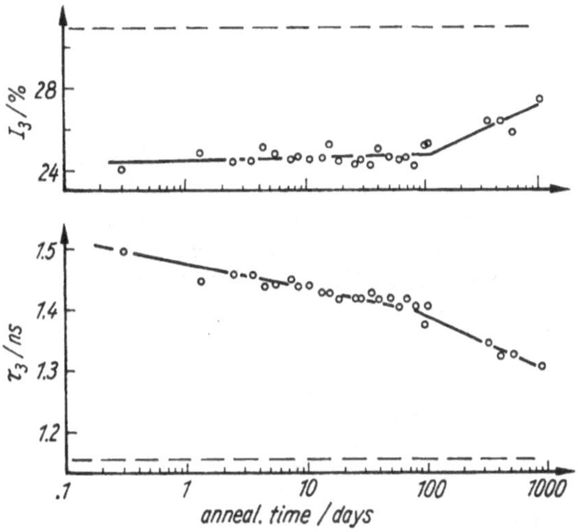

Fig. 4. Time dependence of τ_3 and I_3 following resolidification of AA sample from the melt. Sample was kept at room temperature. Dashed lines represent pre-melt values

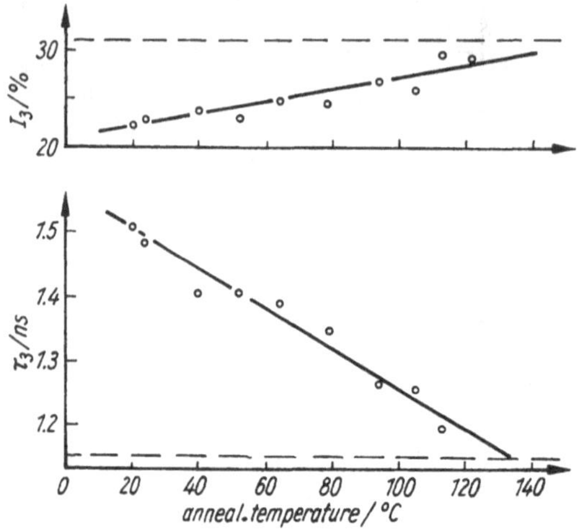

Fig. 5. Dependence of τ_3 and I_3 on annealing temperature for AA sample which had been resolidified from the melt. Dashed lines represent premelt values

4. Discussion

KATAYAMA and HAYNES et al., using ESR, have reported on the free radicals produced by irradiation of AA, but no measurements on the long term stability of these radicals have been reported. In the present work it has been established that positron-detectable effects remain in an X-irradiated sample of AA long after ESR-detectable radicals have decayed. This leads to the conclusion that other products of the irradiation process such as structural changes or lattice defects are also playing a sizeable role in the behaviour of the positron lifetime parameters.

The fact that irradiation by X-rays and resolidification from the melt produce similar effects, i.e., increase of τ_3 and decrease of I_3, might indicate that some similarity exists in the causes of these effects. However, the quite dissimilar behaviour of τ_3 as a function of annealing temperature for the two cases (compare Figs 3 and 5) indicates that the underlying causes are different. Both effects, of course, could be caused by defects but probably different types of defects.

The unusual behaviour of τ_3 for the irradiated sample as the annealing temperature is increased (see Fig. 3) may be related to the lattice expansion observed by KURODA et al. Since these investigators found that the expansion was greatest when assisted by 80 °C heat treatment, the initial increase of τ_3 and I_3 might be explained by the increasing free volume as the irradiated sample is heated. The sudden decrease of τ_3 around 90 °C may occur because a different effect becomes dominant above that temperature.

The fact that heat treatment of AA at 124 °C almost restores the positron lifetime parameters to the pre-irradiation or pre-melt values is not surprising, since annealing at temperatures close to the melting point could eliminate any radicals which might be present or remove other defects caused by either irradiation or resolidification. More experimental information is needed to determine the nature of these defects.

The assistance of Dr. Ichiro Miyagawa in performing the ESR measurements is gratefully acknowledged.

References

ELDRUP, M., LUND-THOMSEN, E., MOGENSEN, O.: J. Chem. Phys. **56** (1972) 4902
HADLEY, J. H., HSU, F. H.: Chem. Phys. Letters **12** (1971) 291
HAYNES, J. C., KURODA, S., MATSUKI, K., MIYAGAWA, I.: Rad. Res. **84** (1980) 426

Ito, Y., Tabata, Y.: Bull. Chem. Soc. Japan **48** (1975) 808
Jean, Y., Ache, H. J.: J. Phys. Chem. **82** (1978) 656
Katayama, M.: J. Mol. Spect. **9** (1962) 429
Kirkegaard, P., Eldrup, M.: Comp. Phys. Commun. **7** (1974) 401
Kuroda, S., Kurita, Y., Miyagawa, I.: Rad. Res. **104** (1985) 116
Walker, W. W., Mueller, E. L.: Appl. Physics **3** (1974) 155

(Received June 16, 1987)

Author's address:
Dr. William W. Walker
Department of Physics & Astronomy
The University of Alabama
Tuscaloosa, Alabama 35487, USA

Cryst. Res. Technol.	23	1988	2	285–290

K. SÜVEGH, CS. SZELES, M. L. WHITE, A. VÉRTES, H. LEIDHEISTER, Jr.

Laboratory of Nuclear Chemistry, Eötvös Loránd University,
Budapest, Hungary, and
Department of Chemistry and Zettlemoyer Center for Surface Studies Lehigh University,
Bethlehem, Pennsylvania, U.S.A.

Positron Lifetime Studies of Organic coatings[1])

Positron implantation and annihilation and the influence of water exposure on the annihilation properties were studied in a number of organic coatings deposited on metal substrates. Change of the lifetime spectra upon water exposure was related to the corrosion protective properties of the coatings.

1. Introduction

The paper is a short summary of our positron lifetime studies on several corrosion protective organic coatings deposited on metal substrates. The work was centered around three main problems of the behaviour and interactions of positrons injected to organic coatings. These are 1) the positron depth distribution in the coating-substrate system and its effect on the measured lifetime parameters, 2) the interpretation of different lifetime components and corresponding positron states, and 3) the sensitivity of positrons to water uptake of the coatings, and its relation to the corrosion protective properties of the coatings. Detailed discussion of the problems were given elsewhere (LEIDHEISER et al., SÜVEGH et al., SZELES et al.).

2. Experimental

Coatings were generally prepared by drow down on cold-rolled steel or iron substrates. Details on the preparation of coatings and the lifetime measurements could be find in the references.

3. Results and discussion

Positron implantation

Positron implantation was studied in two organic coatings, an epoxy and an alkyd paint (SÜVEGH et al.). Positron lifetime spectra were recorded on coatings of varying thickness in the common sandwich geometry. The behaviour of the mean lifetime is given in Figure 1. Since the measured lifetime spectra were superpositions of spectra characterizing the coating and the metal substrate the fraction of positrons transmit-

[1]) Contributed paper presented during the European Meeting on Positron Studies of Defects (PSD 87), March 23–27, 1987, Wernigerode, GDR.

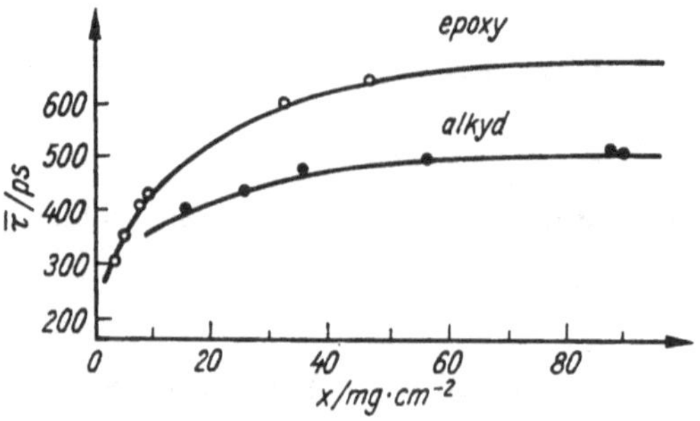

Fig. 1. Positron mean lifetime versus coating thickness. Full curves are the best exponential fits to the data points

ting the coatings were given as

$$f_m(x) = \frac{A(x) - A_p}{A_m - A_p},$$

where $A(x)$ is a measured linear annihilation parameter, while A_p and A_m are the characteristic values of A in the polymer and metal, respectively. We used mean lifetime $\bar{\tau}$, I_1 and I_3 values to calculate transmitted fractions. $f_m(x)$ values showed nearly exponential behaviour common to such experiments. The effective mass absorption coefficient is then

$$\alpha = \frac{d \ln f_m(x)}{dx},$$

while the positron implantation profile to a good approximation is

$$P(x) = -\frac{1}{1-B} \frac{df_m(x)}{dx},$$

were B is the positron backscatter coefficient at the polymer-metal interface. Table 1 shows the extracted effective absorption coefficients, while Figure 2. shows the implantation profile of positrons.

Table 1

Positron absorption coefficients in polymeric coating

Material	Z	α/cm²/g	$\alpha \cdot \varrho$/cm⁻¹
epoxy	2.8	43.2 ± 1.8	48.8 ± 2.0
alkyd	3.0	42.8 ± 2.8	86.9 ± 5.7

Annihilation properties

Positron lifetime spectra of organic coatings in most cases could be resolved into three components and were generally quite similar to spectra recorded on bulk polymers. The short-lived component is in this case however a complicated sum of several contributions: decay of delocalized positrons, para-positronium, and fast pick-off annihilation of both para- and ortho-positronium. In the case of thinner coatings positrons stopped in the metal substrate give their main contribution to this component. The medium-lived component could be associated with annihilation in small

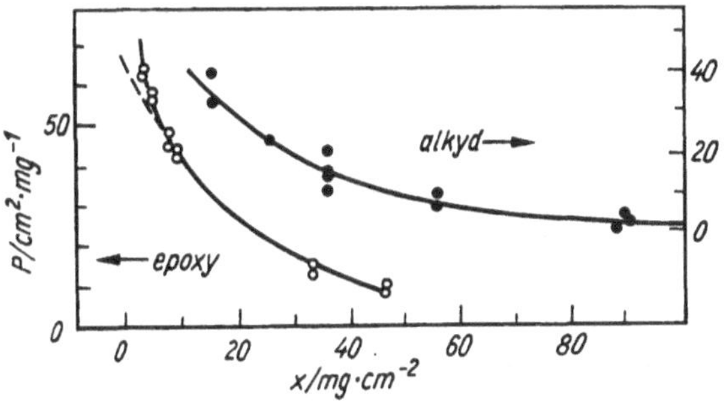

Fig. 2. Positron implantation profiles in alkyd and epoxy coatings. We used the value $B = 0.338$ for the backscatter coefficient of positrons at the polymer-steel interface (MACKENZIE et al.)

voids in the coating; while long-lived component could be associated with decay of ortho-positronium within open volumes in the polymer and at the polymer-metal interface.

Positron lifetime parameters of epoxy and alkyd coatings behave quite similarly with increasing coating thickness, however two main differences appear. Firstly the lifetime parameters of alkyd coatings reach saturation at lower thickness value due to the higher density of this material. Secondly the longest lifetime τ_3 increases strongly with coating thickness in the case of alkyd, while it is essentially constant for epoxy (compare Figs. 3. and 4.). The effect was interpreted by inhomogeneous free volume distribution in alkyd coatings (SZELES et al. 2).

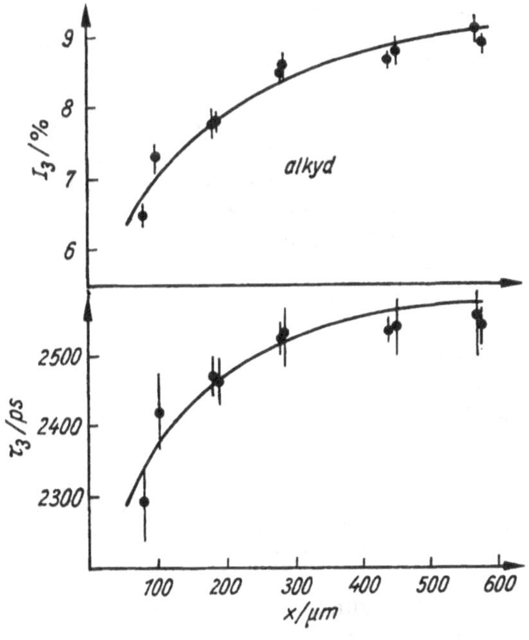

Fig. 3. Intensity I_3 and lifetime τ_3 of the long lifetime component versus coating thickness in alkyd coatings. Full curves show the results of the model calculation using the extracted effective positron absorption coefficient (SZELES et al. 2)

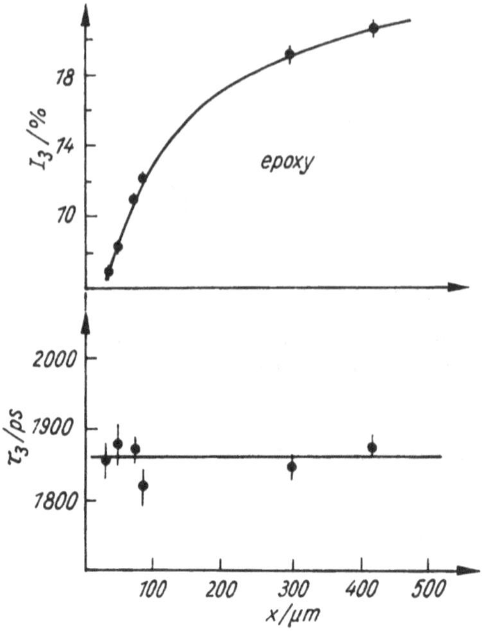

Fig. 4. Intensity I_3 and lifetime τ_3 of the long lifetime component versus coating thickness in epoxy coatings. Full curve for I_3 shows the result of the model calculation using the extracted effective positron absorption coefficient (SZELES et al. 2)

Sensitivity of positrons to water uptake of polymeric coatings

Influence of water exposure on the annihilation properties of positrons injected to the coatings was studied in several corrosion protective polymeric coatings. Positron lifetime spectra were recorded on coatings exposed to room atmosphere denoted with term "dry", and while they were exposed to water denoted with term "wet". The first five coatings in Table 2 were exposed to water for 24 hours and continuously exposed to liquid water at room temperature during the lifetime measurements (LEIDHEISER et al.). The last four coatings were exposed to boiling water for 1 hour and the lifetime measurements were followed at room atmosphere (SZELES et al. 1). The results were compared with the corrosion protective properties of the coatings in an aggressive H_2SO_4 environment. Positron lifetime spectra of coatings providing good corrosion protection and resistance in the acid showed minor changes upon exposure to water. The spectra of coatings showing poor corrosion resistance on the other hand exhibited significant decrease of τ_1 and increase of I_2 on the account of I_1 on exposure to water. The effect could be ascribed to the increase of the trapping probability of positrons in open volumes at the polymer structure due to water uptake (SZELES et al. 1).

References

LEIDHEISER, H. Jr., SZELES, CS., VERTES, A.: Nucl. Inst. Meth. **A255** (1987) 606

MACKENZIE, I. K., SCHULTE, C. W., JACKMAN, T., CAMPBELL, J. L.: Phys. Rev. **A7** (1973) 135

SÜVEGH, K., SZELES, CS., VÉRTES, A., WHITE, M. L., LEIDHEISER, H. Jr.: J. Radional. Nucl. Chem. Lett. **117** (3) (1987) 183

SZELES, CS., VÉRTES, A., WHITE, M. L., LEIDHEISER, H. Jr.: Nucl. Inst. Meth. accepted for publication 1)

SZELES, CS., SÜVEGH, K., VÉRTES, A., WHITE, M. L., LEIDHEISER, H. Jr.: (to be published 2)

(Received June 16, 1987)

Table 2
Influence of water exposure on the annihilation properties of positrons in polymeric coatings

Name	state	thickness μm	lifetime/ps			intensity/%			description of behavior in H_2SO_4
			τ_1	τ_2	τ_3	I_1	I_2	I_3	
vinyl ester	dry	300	195 ± 11	442 ± 18	1671 ± 23	36.6 ± 3.8	46.0 ± 3.5	17.4 ± 0.5	excellent
	wet		202 ± 9	454 ± 17	1666 ± 21	38.4 ± 3.3	44.3 ± 3.0	17.3 ± 0.4	
epoxy	dry	275	198 ± 8	470 ± 14	1679 ± 29	44.8 ± 2.6	44.0 ± 2.3	11.2 ± 0.4	excellent
	wet		180 ± 7	455 ± 13	1719 ± 35	42.2 ± 2.5	48.1 ± 2.2	9.7 ± 0.4	
epoxy amine	dry	375	200 ± 11	440 ± 16	1657 ± 17	33.8 ± 3.5	45.5 ± 3.2	20.7 ± 0.4	good
	wet		191 ± 9	449 ± 12	1710 ± 15	30.9 ± 2.6	48.4 ± 2.3	20.6 ± 0.4	
epoxy I	dry	250	206 ± 10	476 ± 27	1886 ± 35	48.3 ± 4.1	36.5 ± 3.8	15.2 ± 0.5	poor
	wet		178 ± 6	450 ± 11	1820 ± 20	39.3 ± 2.0	45.1 ± 1.8	15.6 ± 0.3	
epoxy Novolac	dry	250	193 ± 7	484 ± 17	1760 ± 24	44.7 ± 2.4	38.6 ± 2.1	16.7 ± 0.4	poor
	wet		161 ± 12	420 ± 13	1741 ± 23	29.9 ± 2.8	51.7 ± 2.5	18.4 ± 0.4	
Pangolin	dry	80	163 ± 3	381 ± 4	2015 ± 167	48.2 ± 1.6	51.1 ± 1.5	0.7 ± 0.2	good
	wet		159 ± 3	370 ± 4	1949 ± 160	45.6 ± 1.3	53.6 ± 1.3	0.8 ± 0.1	
Pavolin	dry	188	178 ± 3	407 ± 9	2507 ± 39	51.7 ± 1.7	37.8 ± 1.7	10.5 ± 0.2	excellent
	wet		177 ± 3	410 ± 8	2329 ± 31	52.8 ± 1.6	38.4 ± 1.6	8.8 ± 0.1	
Orkan	dry	138	169 ± 3	404 ± 5	2316 ± 35	50.9 ± 1.3	45.7 ± 1.2	3.4 ± 0.1	poor
	wet		158 ± 5	393 ± 5	2247 ± 34	47.1 ± 1.2	49.4 ± 1.2	3.5 ± 0.1	
Koralkyd	dry	550	187 ± 9	386 ± 9	2357 ± 20	42.6 ± 2.7	48.6 ± 2.6	8.7 ± 0.1	poor
	wet		164 ± 4	385 ± 5	2282 ± 30	38.0 ± 1.7	53.5 ± 1.6	8.5 ± 0.1	

Authors' addresses:

K. Süvegh, Cs. Szeles, A. Vértes
Eötvös Loránd University, Laboratory of Nuclear Chemistry
H-1088 Budapest, Puskin u. 11—13., Hungary

M. L. White, H. Leidheiser, Jr.
Lehigh University, Department of Chemistry and Zettlemoyer Center
for Surface Studies
Bethlehem, Pennsylvania 18015, USA

Cryst. Res. Technol.	23	1988	2	291–295

J. Ch. Abbé, M. Dębowska

Centre de Recherches Nucléaires, Strasbourg,
and Institute of Experimental Physics, University of Wrocław, Wrocław

Positron Lifetime in Mordenite [1])

Positron lifetime spectra have been measured for mordenite pure and impregnated with various amounts of metallic gold being highly dispersed. The change of annihilation characteristics (in comparison to those corresponding to the pure mordenite), resulting both from the presence of gold particles in the zeolite matrix and the way of treatment of the zeolite during the impregnation, has been noticed.

Получено спектры времен жизни позитронов для образцов чистого морденита и морденита содержащего разные количества сильно диспергированного металлического золота. Обнаружено влияние на аннигиляционные характеристики так самого присутствия золота в матрице зеолита как и условий обработки применяемой в процессе пропитывания зеолита золотом. Возможно что изменение кислотности зеолита, связано с обработкой, затрудняет определение влияния самого присутствия металла в зеолите на аннигиляционные характеристики.

1. Introduction

Zeolites are synthetic and natural aluminosilicates of alkaline and alkaline earth metals whose interesting properties are worthy to be studied. Various forms of synthetic zeolites are very often used in catalysis. Papers dealing with the positron annihilation in zeolites and published up to now contain results sometimes obscure and non-univocal. However the interesting properties of zeolites, such as well-known structure, high specific surface areas, high porosity and variety of their modification, still make them attractive subjects of positron annihilation studies. Metal-loaded zeolites are known for years giving possibility to study the properties of metals of nearly atomic dispersion (Gallezot). In our case the mordenite has been chosen as a matrix for small gold particles. The general information on pure mordenite is given in Table 1 and Figure 1, 2, and 3.

2. Impregnated zeolites

Three gold-loaded zeolites have been produced by the method "to incipient wetness". The details are given in Table 2.

One of the impregnated samples seen by electron microscope is shown in Figure 4.

---

[1]) Contributed paper presented during the European Meeting on Positron Studies of Defects (PSD 87), March 23–27, 1987, Wernigerode, G.D.R.

Electron microscope studies have been performed in the Institute of Low Temperature and Structure Research of Polish Academy of Sciences, in Wrocław.

Table 1

Characteristics of pure mordenite

Formula of ideal unit cell manufacturer	$Na_8[(AlO_2)_8 (SiO_2)_{40}]$ Institut Français du Petrole
atomic ratio Si/Al	5.5
Na content wt%	5.6
symmetry	orthorhombic
unit cell constants/A	$a = 18.13$, $b = 20.49$, $c = 7.52$
fundamental building unit	chain consisting 5-rings of SiO_4 tetrahedra and single AlO_4 tetrahedron (cf. Fig. 1)
channel system	two-dimensional, main system ‖ to c, channels with 2.8 Å restrictions ‖ to b
void volume/cc/cc	0.28

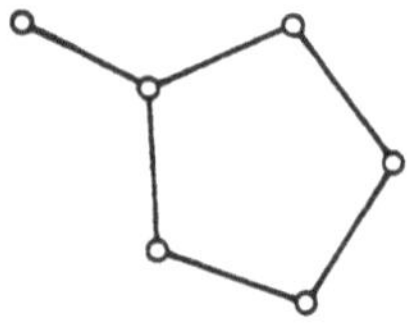

Fig. 1. The fundamental building unit of mordenite

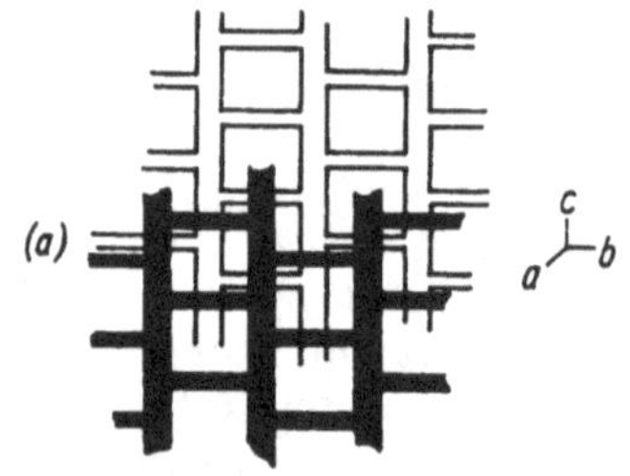

Fig. 2. Channel system n mordenite

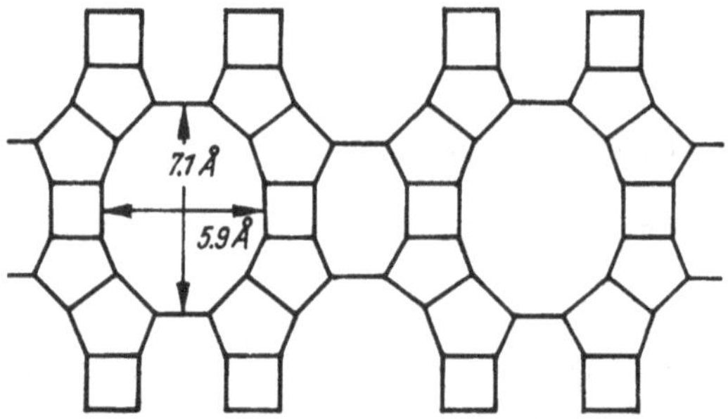

Fig. 3. Pores in mordenite

Table 2

Impregnation of zeolites — successive stages

Activation of pure mordenite	in air at 573 K, 40 hrs		
impregnation with	aqueous solution of $HAuCl_4$		
reduction of $AuCl_4^-$ to Au	in a slow flow of H_2 at 503 K*, 18 hrs		
dispersion of Au	gold content/wt%		
	0.8	2.1	2.5
mean arithmetic diameter of Au particles [Å]	102	89	90
median of Au particles [Å]	66	66	62
location of Au	external surface of mordenite grains		

* This temperature was suggested (ROMANOWSKI) as sufficient for reduction and not too high to cause gold conglomeration.

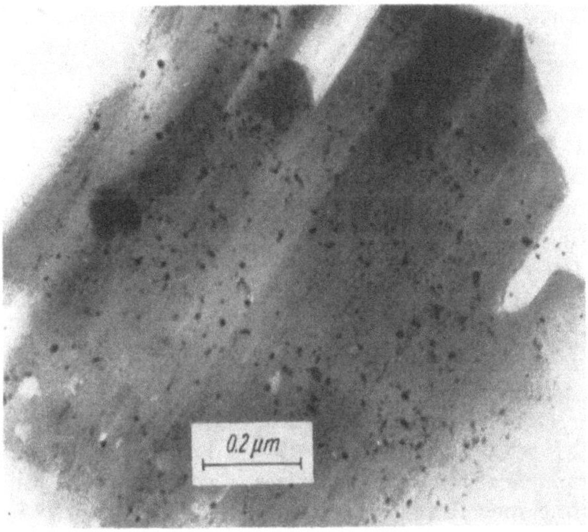

Fig. 4. Transmission picture of Au/mordenite sample obtained by the electron microscope Tesla BS613. Black dots denote gold particles on the external surface of mordenite grains

3. Lifetime measurements

Positron lifetimes have been measured in vacuum of 10^{-4} Pa with use of conventional fast-fast coincidence technique, in Strasbourg. The lifetime spectra have been resolved into three or four components with use of Positronfit Extended programme. The lifetime resolution was about 300 ps. The further information on treatment of the samples before measurements and results of the analysis of the positron lifetime spectra are given in Table 3.

Table 3
Lifetime characteristics

Sample	treatment before measurements	lifetimes/10^{-9} s			intensities/%		
		τ_2	τ_3	τ_4	$I_1 + I_2$	I_3	I_4
I-pure	vacuum heating: 573 K, 17 hrs	0.73	2.27	52.69	77.7	19.7	2.6
II-pure	heating in H_2: 473 K, 18 hrs then very long heating in air/573 K/then heating in vacuum: 573 K, 17 hrs	0.72	2.32	48.50	78.7	19.4	2.0
III-pure	heating in H_2: 503 K, 18 hrs then heating in vacuum: 503 K, 17 hrs	0.71	3.10	—	70.7	29.3	—
mordenite impr.	activation and re-						
0.8%	duction as in Table	0.73	3.13	—	76.3	23.7	—
2.1%	II then heating in	0.67	2.84	—	79.4	20.1	—
2.5%	vacuum: 503 K, 17 hrs	0.69	2.96	—	79.4	20.6	—

4. Conclusions

The procedure of reduction brought about good results as far as the dispersion of gold is concerned; yet it is difficult to define the influence of the metal on the annihilation characteristics. The long-lived components of lifetimes τ_3 and τ_4 prove the formation of positronium in mordenite that has been observed already (Dębowska et al.). It is assumed that:

a) the sum: $I_1 + I_2$ includes annihilation events of positrons in the zeolite framework and annihilation of parapositronium

b) components of τ_3 and τ_4 come from the pick-off annihilation of ortho-positronium in channels of mordenite. It isn't excluded that the two different values of lifetimes reflect the existence of channels of two different sizes in the zeolite (cf Fig. 2).

c) comparison of the lifetime spectra for samples of pure mordenite N⁰ I, II, and III indicates that the temperature of heating of samples, just before the lifetime measurements, has a crucial influence on measured characteristics. The temperature of heating as well as the hydrogen used during reduction could change the acidity of samples (Gallezot) and influence the lifetime characteristics.

d) although because of c) it is difficult to define the real influence of dispersed gold on positron and positronium interaction with the mordenite, it is seen (comparison between spectra for impregnated samples and for mordenite pure N⁰ III) that the dispersed gold augments the sum: $I_1 + I_2$ and reduces the value of I_3. It is consistent with our expectations that positrons should be trapped at gold surface in case of impregnated samples. In preliminary Doppler measurements the increase in a very broad component and in a broad one was observed (Abbé, Dębowska). The last is consistent with results of Mogensen et al. for Fe/SiO_2 and Ewertowski et al. for Pt/Al_2O_3.

This work was partially supported by the Jagiellonian University within the research programme CPBP 01.09.

References

ABBÉ, J. CH., DĘBOWSKA, M.: phys. stat. sol. (b) **146** (1988) 95
DĘBOWSKA, M., BARANOWSKI, A., JERIE, K., ŚWIĄTKOWSKI, W.: phys. stat. sol. (a) **97** (1986) K35
EWERTOWSKI, F., ROZENFELD, B., KLIMAS, J.: Ann. Soc. Chim. Polonorum **47** (1973) 807
GALLEZOT, P.: Catal. Rev.-Sci. Eng. **20** (1979) 121
MOGENSEN, O., ELDRUP, M., MØRUP, S., ØRNBO, J. W., TØPSOE, H.: in Proc. 7th International Conference on Positron Annihilation, eds P. C. JAIN, R. M. SINGRU, K. P. GOPINATHAN, Singapore 1985
ROMANOWSKI, W.: private communication

(Received June 16, 1987)

Authors' addresses:

Prof. J. CH. ABBÉ
Centre de Recherches Nucléaires, Département de Chimie Nucléaire
B.P. 20
67037 Strasbourg Cedex, France

Dr. M. DĘBOWSKA
Institute of Experimental Physics, University of Wrocław
Cybulskiego 36
50-205 Wrocław, Poland

Cryst. Res. Technol.	23	1988	2	K29–K32

Short Notes

L. A. BOROVINSKY, R. N. SHEFTAL

Novgorod Pedagogical Institute, Novgorod, USSR, and
Institute for Radioengineering and Electronics, USSR,
Academy of Siences, Moscow, USSR

Equilibrium Condition and Work of Formation of Nuclei of a Monocrystal Film in a Plane Capillary

In order to get a smooth surface during recrystallization of films and to preserve the original composition for the case of recrystallization of a compound containing a volatile component, a special recrystallization process under a protecting coating is used (ZYBTSEV, SHEFTAL). Thus, substrate and coating form a plane capillary. Equilibrium condition between solid and melt as well as kinetics of formation of crystal nuclei in a plane capillary of about 10^{-7} m thickness are substantially different from a condition of equilibrium and the kinetics of film nucleation on a single substrate. So it is necessary to develop a theory of crystallization in a plane capillary that would take into account peculiar features of such a process. In this article basic propositions of such a theory are considered, i.e. equilibrium condition for solid and melt in a plane capillary is established and expressions for the work of formation of nuclei of a monocrystal film are obtained.

Let us denote parameters belonging to a monocrystal film, substrate, and melt with subscripts "1", "2", "3", respectively. Accordingly, σ_{12} and σ_{23} are specific adhesion energies for the pairs: crystal film-substrate and melt-substrate. σ_{13} being the specific surface energy for a crystal-melt boundary. Let $\Delta\mu$ be the difference of free energies of liquid and solid phases per unit volume of the solid phase. We will assume that the substrate and the protecting coating are of the same material.

To establish the condition of equilibrium for the film and the melt we will consider the free energy for the formation of a film area being in contact with the surfaces of the substrate and the protecting coating and on its side boundaries surrounded by melt. When finding the equilibrium condition we may assume that the dimensions of the chosen film area in the longitudinal direction are macroscopic, that is, much greater than the film thickness. Due to this in the expression for the free energy one can neglect the term corresponding to the surface energy of the side boundaries between the film and the melt. Let h denote the gap thickness and S_1 the total surface of the boundaries for the film area with the substrate and coating. Free energy for the formation of this area of a monocrystal film from melt may be written as follows:

$$F = -\Delta\mu \frac{S_1 h}{2} - (\sigma_{12} - \sigma_{23}) S_1 . \tag{1}$$

From $\frac{dF}{dS_1} = 0$ we find the equilibrium condition:

$$\Delta\mu h = 2(\sigma_{23} - \sigma_{12}) . \tag{2}$$

Thus, it is seen from (2) that if the adhesion energy of crystal and substrate is greater than the adhesion energy of melt and substrate ($\sigma_{12} > \sigma_{23}$), then $\Delta\mu < 0$ i.e. the equilibrium state of film and melt is reached at a temperature that corresponds to the superheated condition of the bulk crystal. If, on the other hand, $\sigma_{23} > \sigma_{12}$, then $\Delta\mu > 0$, i.e. equilibrium is reached at a temperature corresponding to supercooling of the macroscopic melt.

Further we will consider the work of formation of crystal nucleus in a melt in plane capillary. If the gap is substantially greater than the dimensions of the critical nucleus, it is evident that nucleation is going on independently on both surfaces of the capillary and thus may be described by wellknown theories of heterogeneous nucleation (HIRTH, MOAZED). But when the supercooling is not big, the dimensions of the critical nucleus are comparable with the gap width. In this case the existence of two surfaces leads to peculiar features of the nucleation process, substantially different from the above mentioned. Let the crystal nucleus belong to cubic symmetry, its faces (001) lie on capillary's surfaces and let faces (111) be in contact with the melt. A nucleus that fits to one of the capillary's surfaces has the form of a regular tetragonal pyramid. A nucleus that fits to both surfaces has either the form of a truncated pyramid or a bipyramid (see Figs 1a, b, c). Common base of the two pyramids making a bipyramid is the surface of twinning. Its area is smaller than the area of either face fitting to capillary's surfaces.

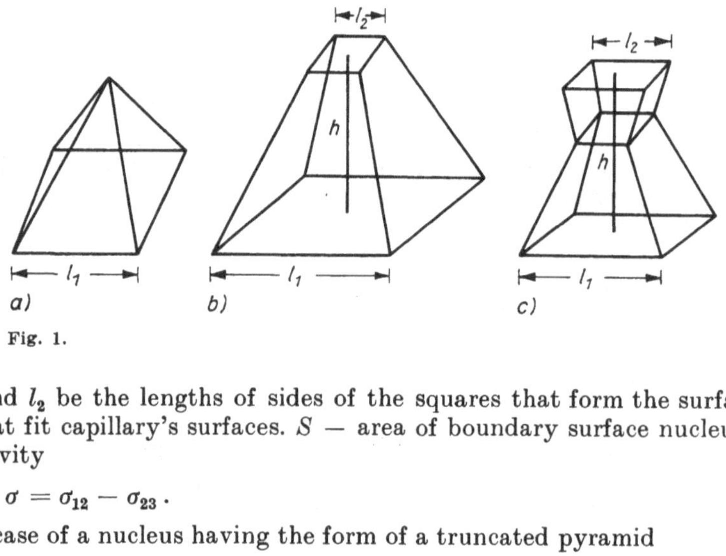

Fig. 1.

Let l_1 and l_2 be the lengths of sides of the squares that form the surfaces of the nucleus that fit capillary's surfaces. S — area of boundary surface nucleus — melt. Let for brevity

$$\sigma = \sigma_{12} - \sigma_{23} \, . \tag{3}$$

For the case of a nucleus having the form of a truncated pyramid

$$|l_1 - l_2| = \sqrt{2}\, h \, . \tag{4}$$

For the case of a nucleus having the form of a bipyramid, l_1 and l_2 are independent geometrical variables, but their difference is limited by the condition

$$|l_1 - l_2| < \sqrt{2}\, h \tag{5}$$

A nucleus having the form of a truncated pyramid may be considered as a limiting case of a nucleus having the form of a bipyramid when the twinning surface coincides with one of capillary's surfaces.

The free energy of nucleus formation is written as follows :

$$F = -\Delta\mu V + \sigma_{13} S - \sigma S_1 \, . \tag{6}$$

Substituting for V, S, and S_1 their expressions in the case of nuclei having the form of a pyramid, a truncated pyramid, and a bipyramid, one obtains

$$F_1 = -\frac{\sqrt{2}}{6}\Delta\mu l_1^3 + l_1^2(\sqrt{3}\,\sigma_{13} - \sigma)\,, \tag{7a}$$

$$F_2 = -\Delta\mu(l_1^2h - \sqrt{2}\,l_1h^2 + \tfrac{2}{3}h^3) + {} + 2\sqrt{3}\,\sigma_{13}h(\sqrt{2}\,l_1 - h) - 2\sigma(l_1^2 - \sqrt{2}\,l_1h + h^2)\,, \tag{7b}$$

$$F_3 = -\Delta\mu\left\{\frac{\sqrt{2}}{8}(l_1^3 + l_2^3 - l_1^2l_2 - l_1l_2^2) + \frac{1}{4}h(l_1 + l_2)^2 - {} - \frac{\sqrt{2}}{4}(l_1 + l_2)\,h^2 + \frac{1}{6}h^3\right\} + \sqrt{3}\,\sigma_{13}\left\{\frac{1}{2}(l_1 - l_2)^2 + {} + \sqrt{2}\,h(l_1 + l_2) - h^2\right\} - \sigma(l_1^2 + l_2^2)\,. \tag{7c}$$

As the height of a nucleus — having the form of a pyramid — is equal to $\frac{1}{\sqrt{2}}l_1$ and does not exceed the gap width h, the expression (7a) is valid under the condition

$$l_1 \leqq \sqrt{2}\,h\,. \tag{8}$$

The sizes of a critical nucleus may be found as stationary points for the functions 7a, 7b, 7c.

The maximum value of the function (7a) corresponds to l_1 equal to

$$l_{1\mathrm{m}} = \frac{2\sqrt{2}\,(\sqrt{3}\,\sigma_{13} - \sigma)}{\Delta\mu}\,. \tag{9}$$

As l_1 in this case should obey (8), values of $\Delta\mu$ that allow nucleation on one surface, satisfy the condition

$$\Delta\mu h = 2\,(\sqrt{3}\,\sigma_{13} - \sigma)\,. \tag{10}$$

Function (7b) has the maximum value for l_1 equal to

$$l_{1\mathrm{m}} = \frac{\sqrt{2}\,h^2\,\Delta\mu + 2\sqrt{6}\,h\sigma_{13} + 2\sqrt{2}\,h\sigma}{2(\Delta\mu h + 2\sigma)}\,. \tag{11}$$

The saddle point of function (7c) corresponds to a critical nucleus having the form of a symmetrical bipyramid with the lengths of base sides

$$l_{1\mathrm{m}} = l_{2\mathrm{m}} = \frac{\sqrt{2}\,h^2\,\Delta\mu + 4\sqrt{6}\,h\sigma_{13}}{4(\Delta\mu h + 2\sigma)}\,. \tag{12}$$

The work of nucleation may be found as the maximum of functions (7a) or (7b) or else as value of function (7c) for a saddle point. This work of nucleation is expressed by the following formulas:

$$W_1 = \frac{8}{3}\,\frac{(\sqrt{3}\,\sigma_{13} - \sigma)^3}{(\Delta\mu)^2}\,, \tag{13a}$$

$$W_2 = \frac{2h^2}{\Delta\mu h + 2\sigma}\left\{-\frac{1}{12}(\Delta\mu h)^2 + 3\sigma_{13}^2 - \sigma^2 - \frac{2}{3}\Delta\mu h\sigma\right\}, \tag{13b}$$

$$W_3 = \frac{2h^2}{\Delta\mu h + 2\sigma}\left\{-\frac{1}{48}(\Delta\mu h)^2 + 3\sigma_{13}^2 - \sqrt{3}\,\sigma_{13}\sigma - \frac{1}{6}\Delta\mu h\sigma\right\}. \tag{13c}$$

Expressions (13b) and (13c) as well as (11) and (12) show that, if $\sigma > 0$, nucleus formation in a plane capillary is possible not only under conditions of supercooling but also when $\Delta\mu = 0$ or even $\Delta\mu > 0$, i.e. under certain superheating, if

$$\Delta\mu h + 2\sigma > 0 \tag{14}$$

is fulfilled.

This is explained by the fact that the increase of volume free energy during crystallization process under conditions of superheating might be compensated due to stronger adhesion of the crystal film with its substrate. It may be shown that in the case when the height of a nucleus is equal to the gap width, i.e. when $\Delta\mu h = 2(\sqrt{3}\,\sigma_{13} - \sigma)$, expressions (13a) and (13b) coincide.

When $\Delta\mu h < 2(\sqrt{3}\,\sigma_{13} - \sigma)$ the existence of the second surface lowers the nucleation barrier. To find out what form of nucleus is preferable one should compare (13b) and (13c) with different values of $\Delta\mu$. The analysis of the difference $W_3 - W_2$ shows that $W_3 > W_2$ when

$$-2\sigma < \Delta\mu h < 4\sqrt{\sigma}\,(\sqrt{\sqrt{3}\,\sigma_{13}} - \sqrt{\sigma})\,, \tag{15}$$

and $W_3 < W_2$ when

$$4\sqrt{\sigma}\,(\sqrt{\sqrt{3}\,\sigma_{13}} - \sqrt{\sigma}) < \Delta\mu h < 2(\sqrt{3}\,\sigma_{13} - \sigma)\,. \tag{16}$$

This means that under condition (15) the prefered form of the nucleus is a truncated pyramid, whereas under condition (16) it is the form of bipyramid.

Conclusions

Features of crystallization in very narrow plane capillaries differ substantially from features of heterogeneous crystallization from melted phase on a substrate. It is shown that for the case when energy of adhesion of crystal film and substrate is greater than energy of adhesion of melt and substrate, equilibrium takes place when temperature is higher than melting point of bulk crystal, in the opposite case — when temperature is lower than melting point. Expressions for the work of formation of crystal nuclei from melt in plane capillary are obtained for the case of one of the possible orientations of monocrystal film having cubic symmetry.

References

HIRTH, J. P., MOAZED, K. L.: Physics of thin films, v. 4, N-4, London 1967

ZYBTSEV, S. G., SHEFTAL, R. N.: Investigation of recrystallization of thin films of InSb by laser refractometry. The processes of growth and synthesis of semiconductor crystals and films. VII Conference USSR, v. 1, p. 69—79. Novosibirsk 1986

(Received September 1, 1987; accepted September 12, 1987)

Authors' addresses:

Dr. R. N. SHEFTAL

Moscow, GSP-3
prospekt Marx' a, 18
USSR, Institute for Radioengineering and Electronics

Prof. Dr. L. A. BOROVINSKY
Novgorod Pedagogical Institute

Cryst. Res. Technol.	23	1988	2	K33–K37

M. A. CABALLERO, M. GONZÁLEZ-MAÑAS

Dpto. Estructura y Propiedades de los Materiales, Puerto Real, Cádiz, Spain

Morphology of Brazilian Twins in Synthetic Quartz Crystals

1. Introduction

Several authors have carried out studies on surface structures in different materials in order to obtain information about growth mechanisms and associated growth defects. Thus, KOTRU (1973, 1978 (A) (B)) observes quartz microcrystals acting as growth nucleus centers on rhombohedral surfaces; JOSHI et al. (1980a) and JOSHI and SARAF study how guest microcrystals modify the growth fronts on second order prism faces; JOSHI et al. (1980b) carry out a comparative microtopographical study of first order prism faces of natural and cultured quartz; KOTRU and RAINA (1982) study growth structures associated with microtwinning phenomena on first order prism surfaces.

TSINOBER et al. pointed out, for the first time, the presence of Brazil twins on principal positive rhombohedron $\{10\bar{1}1\}$ and negative trigonal prism $\{11\bar{2}0\}$ faces in synthetic quartz crystals.

In the present work the authors carry out a study of $-X$ faces $\{11\bar{2}0\}$ with Brazilians twins which are externally manifested by the presence of two-fold symmetry polyhedra on those surfaces in synthetic quartz crystals. Likewise we have studied the Brazilian twin morphology in Z $\{0001\}$ growth sectors when the individual twin development allowed us its observation on those growth sectors.

2. Experimental details

The Y-cut seed hydrothermally grown quartz crystals used in this work were chosen from among those in which Brazilian twins were present on $-X$ growth surfaces. The growth conditions for those crystals were the following ones: growth temperature 354.2 °C, average temperature gradient 12.8 °C, average pressure 1585 bar. The nutrient was natural quartz from Madagascar and the solvent used NaOH.

The surfaces to be studied were carefully cleaned with ethyl alcohol. The etching on Z-plates was carried out using 40% HF solution during two hours. The $-X$ surfaces and the Z-plates were examined under an Olympus BH-2 microscope.

3. Results and discussion

The quartz crystals grown under the conditions mentioned above have the typical morphology corresponding to the Y-cut seed quartz crystals shown in Figure 1. Some of them exhibit, on $-X$ surfaces, several forms constituted by an ensemble of faces with two-fold symmetry. The dimensions of these polyhedra are very variable ranging between the micron and the centimeter size, but they have always the same orientation: 23° towards the Z-axis [0001] and their morphology is very similar, as it can be seen in Figure 2.

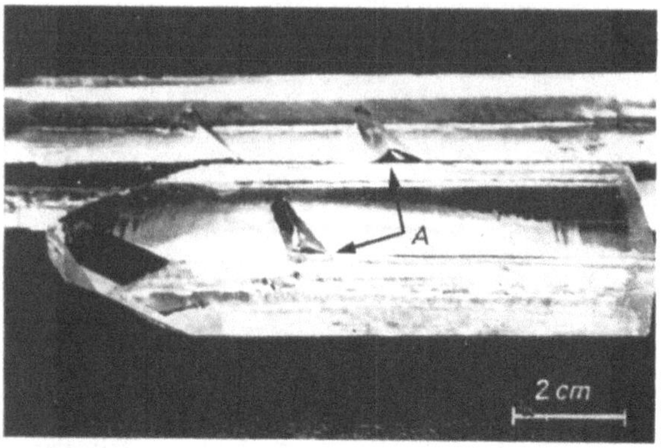

Fig. 1. A characteristic Y-cut seed of hydrothermally grown quartz crystal showing Brazilian twins on −X faces

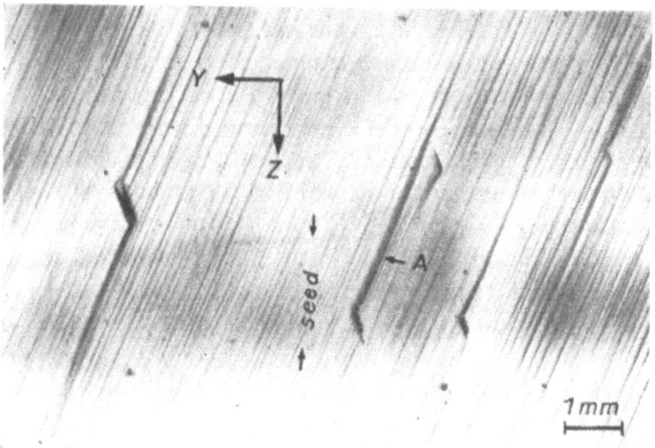

Fig. 2. The distribution and orientation of Brazilian twins on −X striated surfaces

The distribution of these polyhedra on −X surfaces is connected to the seed in such a manner that there is no polyhedron on zones near the −X edges (see Fig. 2). When these forms have a maximum development their morphology on Z surfaces {0001} can be observed being always a "boot-like" shape.

All the above cited points out that the growth of micronuclei generating these Brazil twins can appear at any moment during the crystal growth, but their distribution suggests that the seed affects not only the generation of growth defects, such as dislocations, but also the appearance of Brazil twins, in some way.

A more particular study manifests that those polyhedra are constituted by four faces related by the −X point symmetry, a two-fold axis, as it can be seen in Figure 3. Nevertheless the trapezoidal shape faces, which appear (marked with A in Figure 3) are actually composed by at least two different faces. The Figure 4a, a higher magnification of the zone marked with a square in Figure 3, shows three faces having very different microtopography: the central face (number 2 in the photo) exhibit homogeneous and uniform growth step distribution, their heights are similar all along the surface, whereas on face 3, this distribution is not uniform and the steps are higher than those corresponding to face 2 (see Figs 3 and 4a); in Figure 4b it can be observed an irregular boundary restrained to the region of surface 1 close to the adjacent face. This boundary may be produced by the interaction of growth layers of the two faces.

The −X faces show, either in twinned crystals or in crystals free of this type of defect, a striation system (see Figs 2 and 3) parallel to the Z direction in the margins of the face and making an angle of 28° with this direction in the central region of this face, near to the seed. These striations are, in fact, a phenomenon tied to the growth process but not to the twinning. They cannot be easily assigned to growth striations similar to those appearing frequently on prism faces {10$\bar{1}$0} in natural quartz since in our case the striation system is not perpendicular to the Z direction. In our opinion these striations can be produced by a superficial craking process or, may be, by phenomena of unstable thermal convection flows inside the pressure

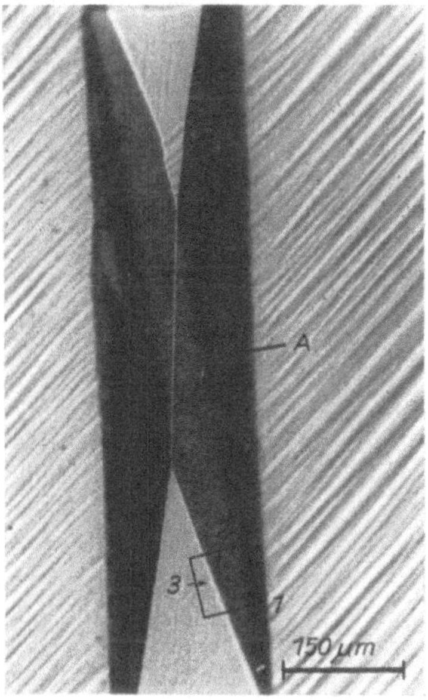

Fig. 3. The typical morphology of Brazilian twins

Fig. 4. (a). A higher magnification of the zone marked with a square in Figure 3 exhibiting three different surfaces and growth steps on the central face;
(b). An irregular boundary in the region of face 1 close to the edge

vessel. In Figure 2, it can be observed the perturbations, marked with A in the photo, generated by the Brazil twins on −X surfaces. It is difficult to stablish the origin of these striations without ambiguity, but the oriented perturbations connected with the striations suggest that the unstable thermal convection flows are on the basis of both defects, striations and Brazilian twins.

The morphology of these Brazil twins on Z faces can be clearly determined examining Z-plates under polarized light (see Fig. 5). This "boot-like" morphology represents two different types of boundaries between right and left individuals: in the −X growth sector the boundaries are straight and parallel, similar to those described by LANG, by PHAKEY and by LANG and MIUSCOV in the rhombohedron region in natural quartz; on the contrary, in the Z growth sector these boundaries show straight and irregular shapes. It is important to point out that the twinned zones never surpass the sub-boundary between the Z und Z' growth sectors.

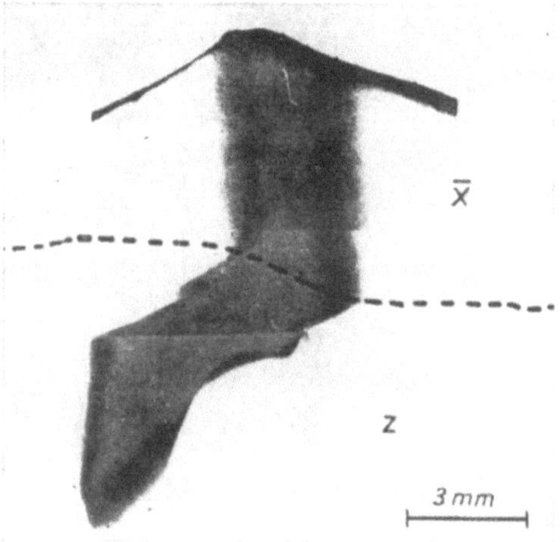

Fig. 5. The typical "boot-like" morphology of Brazilian twins on Z-plates. The straight and irregular boundaries on −X and Z growth sectors can be observed

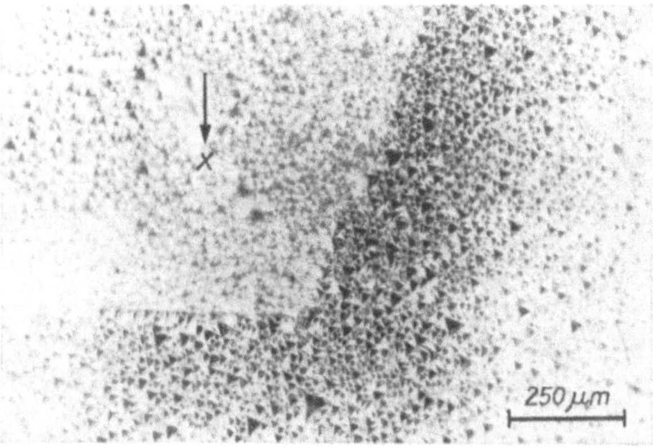

Fig. 6. The etch pits corresponding to right and left individuals on Z-plates

The etching carried out on these plates clearly shows the different orientation of the etch pits in both individuals (see Fig. 6) and how the twinned parts are related by reflection over $\{11\bar{2}0\}$ plane, perpendicular to a two-fold axis.

The presence of Brazilian twins in synthetic quartz crystals generates an ensemble of growth defects, fundamentally planar defects, associated to variations in the growth directions that will be the object of a next publication.

References

JOSHI, M. S., SARAF, K. B., ANTONY, A. V.: Bull. Mater. Sci. **2** (1980) 71
JOSHI, M. S., SARAF, K. B.: Indian J. Pure and Appl. Physics **18** (1980) 162
JOSHI, M. B., KOTRU, P. N., VAGH, A. S.: Kristall und Technik **15** (1980) b, 1003
KOTRU, P. N.: Japanese J. Appl. Physics **12** (1973) 790
KOTRU, P. N. (A): Kristall und Technik **13** (1978) 35
KOTRU, P. N. (B): Indian J. Pure and Appl. Physics **13** (1978) 537
KOTRU, P. N., RAINA, K. K.: Cryst. Res. Technol. **17** (1982) 1077
LANG, A. R.: Appl. Physics Lett. **7** (1965) 168
LANG, A. R., MIUSCOV, V. F.: Growth of Crystals **7** (1969) 112
PHAKEY, P. P.: phys. stat. sol. **34** (1969) 105
TSINOBER, L. I., KHADZI, V. E., GORDIENKO, L. A., SAMOILOVICH, M. I.: Growth of Crystals **6A** (1968) 25

(Received September 4, 1987; accepted September 12, 1987)

Authors' address:
Prof. M. A. CABALLERO, M. GONZÁLEZ-MAÑAS
Dpto Estructura y Propiedades de los Materiales, Facultad de Ciencias
Ap. 40
11510 Puerto Real (Cadiz)
Spain

Cryst. Res. Technol.	23	1988	2	K38–K39

Č. BARTA, Jr., Č. BARTA, Z. PAVLICEK, O. GILAR

TESLA Research Institute of Nuclear Instruments Premysleni, Czechoslovakia, and
Institute of Physics, ČSAV, Prague, Czechoslovakia

Diffusion of Mercury Bromide Gases in Calomel (Hg_2Cl_2) Single Crystals

1. Introduction

Some applications require that a thin layer showing higher refractive index be produced on the surface of an Hg_2X_2 (X = Cl, Br, I) crystal element. The study reported presents basic knowledge and experience with the production of such a layer on calomel (Hg_2Cl_2) single crystals using the closed-system diffusion method at elevated temperature and partial mercury chloride and bromide pressures. A new finding is the respective value of the diffusion coefficient.

2. Experimental procedure

Calomel crystal specimen were ground and polished into a shape of plates of $10 \times 10 \times 2.3$ mm in size. The actual experimental installation consisted of a quartz glass resistance tube furnace in which a sealed quartz ampoule was inserted containing a calomel crystal with a mixture of mercury chloride and mercury bromide powder. Thus, the whole configuration allowed visual inspection throughout the experiment.

The powder mixture inside the ampoule consisted of mercury (I) chloride and mercury (I) bromide in a ratio of 3 : 1. The mercury (I) bromide was the actual diffusion medium while the presence of the fine mercury (I) chloride powder served to minimize thermal etching of the specimen. Because of its large specific surface, the calomel powder changed preferably into the gaseous phase after heating. Thus, the single crystal specimen itself remained practically unetched.

The required operating temperature was estimated by comparing the values of the diffusion coefficients and the partial pressures of comparable materials to be approximately 350 °C. The experiments were carried out at a temperature of 352.5 ± 0.3 °C over 52 hours (two specimens) and 96 hours (two specimens), respectively.

3. Results and discussion

After finishing the experiment the specimens were split in the direction (110) in order that the concentration profile of the diffusing material (Br) could be observed on the surface thus produced, using an electron microprobe. The concentration distribution of the diffused Br is shown in Figures 1 and 2 where the experimental values are plotted with a solid line; the dashed line shows the concentration profile derived theoretically under the assumption that the diffusion coefficient is not concentration dependent. In the present case the mean value of the diffusion coefficient $D = 4.5 \times 10^{-11}$ cm^2 s^{-1}. The comparison of the both curves shows that the experimental curve is, in contrast to the theoretical curve, an "S-shaped". This can be explained by the fact that at higher Br concentrations in the calomel surface layers the diffusion coefficient is concentration dependent.

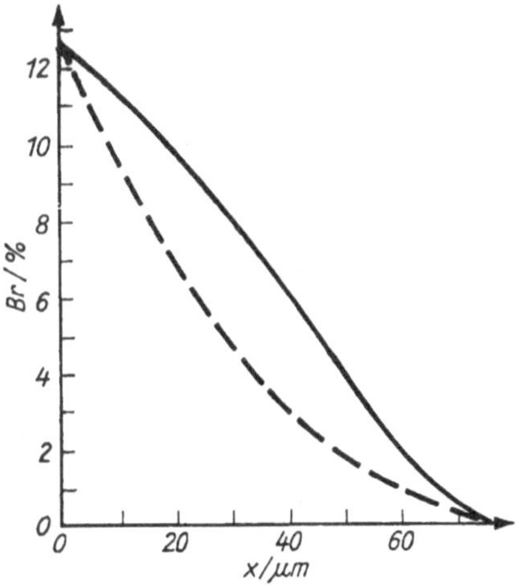

Fig. 1. Concentration distribution of bromide in dependence on the distance from the specimen edge following diffusion heating at a temperature of 352.5 °C over 52 hours. The experimental curve is plotted in solid line, the theoretical curve in dashed line

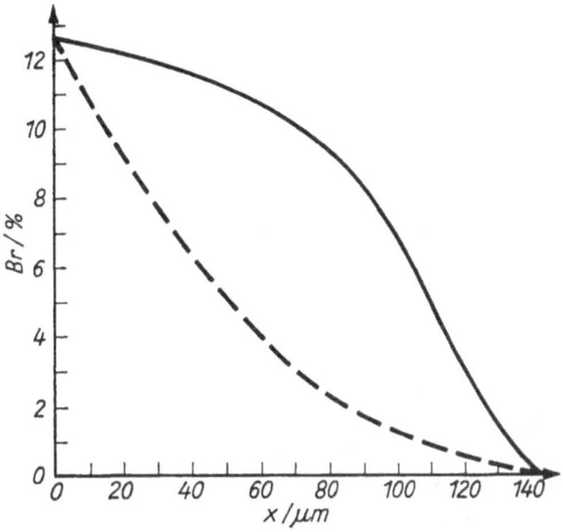

Fig. 2. Concentration distribution of bromide in dependence on the distance from the specimen edge following diffusion heating at a temperature of 352.5 °C over 96 hours. The experimental curve is plotted in solid line, the theoretical curve in dashed line

4. Conclusion

The experiments showed that there is a principle possibility to produce a layer on the surface of calomel the refractive index of which is increased in the given degree (depth). The study of this process is still in progress.

Both, the experience gained and the results obtained, as well as the technology developed, will be used for measurements of the temperature dependence of the diffusion coefficient in the given system. It is expected that the results allow to obtain the basic constants characterizing the temperature dependence of the diffusion coefficient which is described by the equation (Crank)

$$D = D_0 \cdot \exp\left(-\frac{Q}{RT}\right),$$

where

D diffusion coefficient
D_0 frequency factor
Q activation energy
R gas constant.

References

Crank, J.: The Mathematics of Diffusion, Oxford, Second Edition 1975

(Received September 12, 1987; accepted October 2, 1987)

Author's addresses:

C. Barta, Jr., Dr. C. Barta
Institute of Physics ČSAV
Na Slovance 2
180 40 Prague 8, Czechoslovakia

Dr. Z. Pavlicek, Dr. O. Gilar
TESLA Research Institute of Nuclear Instruments
255 66 Premysleni
Czechoslovakia

Cryst. Res. Technol.	23	1988	2	K40–K42

G. KÜHN, B. SCHUMANN

Sektion Chemie, Wissenschaftsbereich Kristallographie, Karl-Marx-Universität Leipzig

$CuInSe_2$ Epitaxial Layers with Sphalerite Structure

Bulk $CuInSe_2$ at 1083 K has a reversible phase transition of first order from the ordered chalcopyrite to the disordered sphalerite structure (MÖLLER et al. 1985). In epitaxial layers deposited on {111}-oriented GaAs and Ge substrates by the flash evaporation technique (SCHUMANN et al. 1978a, 1980a, 1981) the sphalerite phase was observed already at lower substrate temperatures (Tab. 1). This sphalerite structure is stable at room temperature in contrast to the bulk material and can be detected by reflection high energy electron diffraction (RHEED) technique. On {100}- and {110}-oriented substrate no sphalerite structure was found (TEMPEL et al.). It should be noted that parts of the sphalerite phase in the $CuInSe_2$ epitaxial layers are not detectable by RHEED beside the chalcopyrite structure. Therefore, it is possible that the layers consist of the sphalerite phase partially even at lower substrate temperatures as given in Table 1.

Table 1
Experimental results for the epitaxy of flash-evaporated $CuInSe_2$ layers on {111}- oriented substrates

Substrate	range of epitaxial growth T_e/K	structure	references
GaAs {111} A	720–865	chalcopyrite	SCHUMANN et al.
	870	sphalerite	1978a
{111} B	695–820	chalcopyrite	SCHUMANN et al.
	845–870	sphalerite	1981
Ge	720–770	chalcopyrite	SCHUMANN et al.
	795–820		1980a
CaF_2	775–800	chalcopyrite	SCHUMANN,
		(polycryst.	KÜHN
		parts)	

The reason for this phenomenon is unknown so far. In this paper it is tried to give an interpretation for the occurrence of the sphalerite phase in epitaxial layers at relatively low growth temperatures.

Combined thermogravimetric and mass spectroscopic investigations showed that $CuInSe_2$ evaporates in a two-stage process according to

$$2\,CuInSe_2\,(s) \rightarrow Cu_2Se\,(s) + In_2Se\,(g) + Se_2\,(g)$$

(HÖNLE, KÜHN). At lower temperatures (up to about 1360 K) the gaseous phase consists of In_2Se and Se_2. Therefore, these both compounds should occur too at the reevaporation from the growing layers. In accordance with these considerations, at substrate temperatures $T_s \gtrsim 770$ K the epitaxial layers showed a Cu excess in comparison with the stoichiometric composition (HÖBLER et al.; SOMMER et al.).

For the explanation of the existence of the sphalerite phase on the {111} substrate orientation besides other phenomena a face specific desorption process is taken into account. The residence time τ of a molecule on the substrate surface is given by

$$\tau = \frac{1}{\nu} (\exp E_d/k_B T)$$

(GLANG), where ν is the vibrational frequency of adatoms, E_d the desorption energy, k_B the Boltzmann constant, and T the temperature. Both parameters E_d und τ are dependent on the surface structure, as the density of dangling bonds on a surface of a compound with sphalerite or chalcopyrite structure decreases in the following order

$$\{100\}, \{001\} > \{110\}, \{012\} > \{111\}, \{112\} ;$$

for E_d is to expect the reverse tendency.

Considering that the reevaporation from the substrate surface is increasing at higher temperatures and that then the In–Se part compared with Cu is predominant, then a deviation of the layer stoichiometry towards Cu excess will occur firstly at {111} and {112} planes. Furthermore, the epitaxial growth of the sphalerite phase compared to the chalcopyrite phase yields an advantage of energy, because the plane point symmetry groups of deposit and substrate coincide and the type of epitaxy is three-dimensional in comparison with the occurrence of multiple orientation (SCHUMANN et al. 1986). Besides, that is connected with a reduction of crystal defects like dislocations and antiphase domain boundaries.

From the Cu_2Se–In_2Se_3 pseudobinary of the ternary Cu–In–Se system (see, for example, PALATNIK, ROGACHEVA) it follows that the phase δ with sphalerite structure is stable only in the Cu_2Se-rich region contrary to the chalcopyrite phase γ. With the growth of $CuInSe_2$ bulk crystals in this region Cu_2Se precipitates were detected by electron microprobe (MÖLLER et al. 1978). According to the above-mentioned argumentation, in epitaxial layers on {111}-oriented substrates the Cu_2Se-rich δ-phase is stabilized at relatively low growth temperatures. The activation energy for the precipitation from the strained layer probably is not reached.

In epitaxial layers on {111}-oriented GaAs substrates of the In-Se richer compound $CuIn_2Se_{3.5}$ under the same growth conditions, no sphalerite structure could be detected (NEUMANN et al. 1983). RBS measurements have always shown an In-excess compared to Cu (HÜBLER et al.). The same is also valid for the single evaporation technique, in which always an excess of In–Se in the gaseous phase is to be expected (NEUMANN et al. 1980).

Liquid phase epitaxially grown $CuInSe_2$ layers deposited in the same temperature range on {111}-oriented ZnSe substrates have shown the chalcopyrite phase only, because in this process no In–Se depletion can occur (TAKENOSHITA).

It is noteworthy that with flash-evaporated $A^IB^{III}C_2^{VI}$ layers on {111}-oriented CaF_2 substrates the sphalerite phase was not observed (SCHUMANN, KÜHN), though a Cu-excess was measured at higher substrate temperatures (SOMMER et al.). The cause could be, as always polycrystalline parts (differences of the thermal expansion coefficients and distinctly ionic bonds of the substrate) were observed in the films and the strain within the layers is reduced.

It should be noted that the sphalerite phase was observed for other $A^IB^{III}C_2^{VI}$ epitaxial layers deposited on {111}-oriented substrates too: $CuInS_2$ (NEUMANN et al. 1979), $CuGaSe_2$ (SCHUMANN et al. 1978b), Cu(In, Ga)Se_2 (SCHUMANN et al. 1980b).

References

GLANG, R.: in: L. I. MAISSEL, R. GLANG (eds), Handbook of Thin Film Technology, New York 1970
HÖBLER, H. J., FLAGMEYER, R., SCHUMANN, B.: Cryst. Res. Technol. **18** (1983) 61
HÖNLE, W., KÜHN, G.: J. Thermal Analysis **31** (1986) 589
MÖLLER, W., KÜHN, G., BEIER, W.: Kristall u. Technik **13** (1978) 1439
MÖLLER, W., KÜHN, G., ENGEL, W.: J. Thermal Analysis **30** (1985) 383
NEUMANN, H., NOWAK, E., SCHUMANN, B., KÜHN, G.: Thin Solid Films **74** (1980) 197
NEUMANN, H., SCHUMANN, B., NOWAK, E., TEMPEL, A., KÜHN, G.: Cryst. Res. Technol. **18** (1983) 895
NEUMANN, H., SCHUMANN, B., PETERS, D., TEMPEL, A., KÜHN, G.: Kristall u. Technik **14** (1979) 379
PALATNIK, L. S., ROGACHEVA, E. J.: Dokl. Akad. Nauk SSSR **174** (1967) 80
SCHUMANN, B., GEORGI, C., TEMPEL, A., KÜHN, G., NGUYEN VAN NAM, NEUMANN, H.: Thin Solid Films **52** (1978) a, 45
SCHUMANN, B., KÜHN, G.: Cryst. Res. Technol. **19** (1984) 1079
SCHUMANN, B., NEUMANN, H., NOWAK, E., KÜHN, G.: Cryst. Res. Technol. **16** (1981) 675
SCHUMANN, B., NEUMANN, H., TEMPEL, A., KÜHN, G., NOWAK, E.: Kristall u. Technik **15** (1980) b, 71
SCHUMANN, B., TEMPEL, A., GEORGI, C., KÜHN, G.: Thin Solid Films **70** (1980) a, 319
SCHUMANN, B., TEMPEL, A., KÜHN, G.: Solar Cells **16** (1986) 43
SCHUMANN, B., TEMPEL, A., KÜHN, G., NEUMANN, H., NGUYEN VAN NAM, HÄNSEL, T.: Kristall u. Technik **13** (1978) b, 1285
SOMMER, H., JOHN, W., MEISEL, A.: Cryst. Res. Technol. **21** (1986) 595
TAKENOSHITA, H.: Solar Cells **16** (1986) 65
TEMPEL, A., SCHUMANN, B., KOLB, K., KÜHN, G.: J. Cryst. Growth **54** (1981) 534

(Received September 17, 1987; accepted October 2, 1987)

Authors' address:

Prof. Dr. G. KÜHN, Dr. B. SCHUMANN
Karl-Marx-Universität Leipzig
Sektion Chemie, Wissenschaftsbereich Kristallographie
Scharnhorststraße 20
Leipzig
7030
GDR

Cryst. Res. Technol.	23	1988	2	K43–K46

S. A. HODOROWICZ, W. LASOCHA

Department of Crystal Chemistry and Crystal Physics, Jagiellonian University, Kraków, Poland

Structural Characterization of the Ammonium Trimolybdate Hydrate $(NH_4)_2Mo_3O_{10} \cdot H_2O$

When normal molybdate solutions are acidified, various isopolymolybdates of increasing degrees of condensation are formed according to different acidifications and other conditions (SASAKI, SILLEN; TSIGDINOS; TYTKO, GLEMSER). Attempts to identify some complexes have not, however, led to definite conclusions about their existence. In particular, LINDQUIST — basing on an analytical and X-ray powder study — suggested that ammonium trimolybdates $(NH_4)_2O \cdot 3\,MoO_3 \cdot aq$ described by some authors (MELLOR; GUITER; CHANS et al.) may be double salts of hepta- and octamolybdate. Recently, fibrillar crystals of ammonium isopolymolybdate have been isolated (LASOCHA, HODOROWICZ). The results of physicochemical analyses correspond to the formula $(NH_4)_2Mo_3O_{10} \cdot H_2O$. In this paper we focus our attention on the structural characterization of the compound obtained.

1. Experimental

Crystals of $(NH_4)_2\,Mo_3O_{10} \cdot H_2O$ grow only in fibrous form, and the structure determination was infeasible by single crystal methods. The X-ray investigations were crried out on the aggregates of fibres of parallel orientation. They give sharp diffraction spots on rotation photographs taken around the fiber axis. The fibers in the aggregates were oriented by washing in the stream of the mixture of acetone and water. The proper samples about 1–2 mm long and 0.1 mm thick were cut under microscope after drying the aggregates. X-ray photographs were obtained by rotating and moving film methods on a Weissenberg camera. The general set of reflections was recorded on X-ray powder diffractometer DRON-2. Cu-Kα radiation and Ni filter were used. A goniometer was calibrated with a standard sample of quartz. The cell dimensions were refined by the least-squares procedure based on 30 coincidence free reflections. Calculations were carried out on the CDC CYBER computer with the program LATCON (SCHWARZENBACH). The density of crystals was determined by the pycnometric method in CCl_4 under vacuum.

2. Results and conclusions

The synthesis of fibrillar ammonium trimolybdate was carried out using the method described in previous paper (LASOCHA, HODOROWICZ). The most crystals obtained are not single crystals, but consisted of numerous parallelly oriented fibers (Fig. 1). The monocrystals present white fibers a few miillmeters long and about 2–3 μm thick.

From an analysis of the diffraction patterns it was found that the fibrillar ammonium trimolybdate monohydrate crystallizes in the monoclinic system showing the lattice parameters: $a = 9.643(4)$, $b = 7.583(3)$, $c = 8.549(4)$ Å and $\beta = 112.88(3)°$. The density of crystals at temperature 20 ± 0.5 °C is $D_{pyc} = 2.85 \pm 0.01$ g cm^{-3}. The number Z of molecules present in the unit cell of trimolybdate was found to be 2 and the X-ray density $D_X = 2.89$ gcm^{-3}. The systematic absences of $k = 2n + 1$ for $0k0$ and their lack for hkl set made possible the determination of the space groups as $P2_1$ and $P2_1/m$.

The aggregates of crystal fibers of parallel orientation were mounted on a Weissenberg camera and their reflections recorded with exposures ranging up to 60 h. Three films were exposed simultaneously in the multiple film technique. All reflections on Weissenberg photographs showed the shape of sharp lines parallely oriented. Intensities were measured on a microphotometer. The usual Lorentz polarization and absorption corrections for cylindrical shape of sample taken from INTERNATIONAL TABLES were applied.

The structure was solved by means of the heavy atom method. The computations were based on 35 values of $|F_{h0l}|^2$ using the SHELX-76 system of programs. Positions of the molybdenum atoms in the 010 projection were deduced from a Patterson map ($R = 0.25$). Several cycles of the least-squares calculation and the subsequent Fourier

Table 1
Fractional atomic coordinates

Atom	x/a	y/b	z/c	U
Mo 1	0.162(4)	0.0	−0.061(4)	0.04
Mo 2	0.127(7)	0.25	0.209(8)	0.04
0 1	0.138(23)	0.0	0.206(26)	0.08
0 2	0.288(18)	0.25	0.424(20)	0.08
0 3	0.113(24)	0.25	0.462(19)	0.08
0 4	0.163(26)	0.0	−0.236(17)	0.08
0 5	0.329(16)	0.0	0.093(25)	0.08
0 6 or N	0.432(24)	0.0	−0.260(26)	0.08
0 7 or N	0.418(37)	0.25	0.555(44)	0.08

Fig. 1. Growing package of fibers

synthesis ($R = 0.15$) permitted to localize 22 of 26 non-hydrogen atoms. Table 1 and 2 list the atomic coordinates obtained and the observed and calculated structure amplitudes, respectively. A projection of the structure along the b direction is shown in Figure 2 together with the numbering scheme of the atoms.

The unit cell contains of two $[Mo_3O_{10}]^{2-}$ anions, four $[NH_4]^+$ cations and two water molecules; their locations were not elucidated in the present study. The structure is close to that of rubidium trimolybdate $Rb_2Mo_3O_{10} \cdot H_2O$ (KREUSLER et al.). All the

Table 2
Observed and calculated structure factors

hkl	F_0	F_c	*hkl*	F_0	F_c
100	654	643	$\bar{3}04$	609	−456
001	723	800	401	597	−618
$\bar{1}01$	423	504	$\bar{5}03$	582	654
101	329	397	$\bar{2}05$	210	−236
$\bar{2}01$	192	−206	$\bar{5}04$	340	497
200	418	−413	$\bar{4}05$	430	335
$\bar{1}02$	184	−203	204	561	569
002	156	52	$\bar{4}02$	261	355
$\bar{2}02$	700	−616	$\bar{2}03$	901	−863
201	536	−471	$\bar{5}02$	264	353
$\bar{3}01$	438	−512	$\bar{5}01$	217	66
300	961	−949	600	263	391
003	335	−102	$\bar{1}05$	245	−225
301	597	−613	403	318	−220
103	336	322	006	275	−290
400	840	−742	$\bar{3}07$	301	352
$\bar{4}03$	692	532	$\bar{4}06$	354	314
$\bar{2}04$	717	−726			

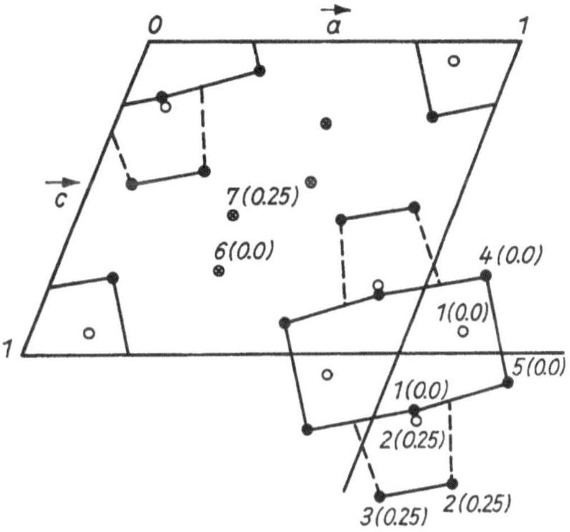

Fig. 2. Projection of the structure on (010). ○ Mo, ● O, ⊗ H_2O or NH_4^+

Mo atoms are octahedrally surrounded by O atoms, and the MoO_6 octahedra form $[Mo_3O_{10}^{2-}]_\infty$ infinite chains. These chains are held together by bridging hydrogen bonds of the types O...H...N and probably O...H...O.

The authors are grateful to the Polish Academy of Sciences for financial support (Project Number: CPBP 01.12).

References

Chans, I. S., Koval, L. B., Sheka, I. A.: Ukr. Khim. Zhurn. **44** (1978) 787
Guiter, H.: Bull. soc. chim. France, **12** (1945) 74
INTERNATIONAL TABLES for X-ray Crystallography, Birmingham, England, Vol. II 1959 and Vol. III 1962

KREUSLER, H. U., FORSTER, A., FUCHS, J.: Z. Naturforsch. **35b** (1980) 242
LASOCHA, W., HODOROWICZ, S. A.: Polish J. Chem. **59** (1985) 971
LINDQUIST, I.: Ark. Kemi **2** (1950) 341
MELLOR, J. W.: A Comprehensive Treatise on Inorganic and Theoretical Chemistry, Vol. XI, London—New York—Toronto 1931, p. 588
SASAKI, Y., SILLEN, L. G.: Ark. Kemi **29** (1968) 253
SCHWARZENBACH, D.: Program for the LS-Refinement of Lattice Constants, University Lausanne, Switzerland 1975
SHELDRICK, G. M.: SHELX-76. Program for Crystal Structure Determination, University of Cambridge, England 1976
TSIGDINOS, G. A.: Bulletin Cdb-12a, Climax Molybdenum Co., Greenwich, Conn. 1969
TYTKO, K. H., GLEMSER, O.: Adv. Inorg. Chem. Radiochem. **19** (1976) 239

(Received October 10, 1987)

Authors' address:

Doc. Dr. habil. S. A. HODOROWICZ, W. LASOCHA
Faculty of Chemistry, Jagiellonian University
Karasia 3
30-060 Kraków,
Poland

Cryst. Res. Technol. | 23 | 1988 | 2 | K47–K50

G. PELZL, C. SCHOLZ, H. SACKMANN

Forschungsabteilung der Sektion Chemie, Martin-Luther-Universität Halle-Wittenberg, DDR

Reentrant Nematic Phases in Binary Systems of Terminal-Nonpolar Compounds (IV)[1])

Binary Systems with Symmetrical Nematogenic Compounds

1. Introduction

Reentrant nematic (N_{re}) phases were observed in a number of binary systems of terminal-nonpolar compounds (PELZL, LATIF et al.; PELZL, SCHOLZ et al.; PELZL, SCHOLZ, DEMUS, SACKMANN; PELZL, DIELE et al.). In these binary systems the one component has a wide nematic range whereas the other component is a smectogenic compound. In all cases studied the nematogenic component possesses an asymmetrical molecular shape. The molecules exhibit in 4,4′ position either two terminal alkyl chains of quite different length or only one long terminal alkyl chain. It was found that a minimum length of the alkyl chains of the nematogenic and smectogenic components is necessary to observe N_{re} phases in such binary systems.

Now we present firstly binary systems of this type where the terminal-nonpolar nematogenic component exhibit two relatively long end-standing alkyl chains.

2. Materials

Nematogenic components:

A:

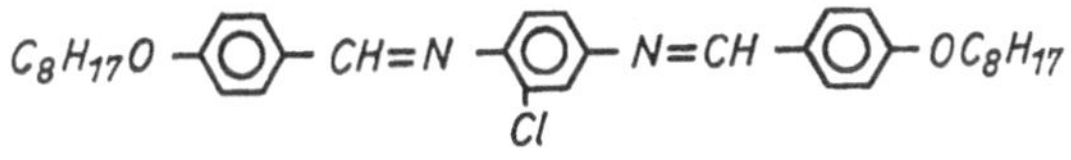

K 59 N 181 Is (NEUBERT, MAURER)

B:

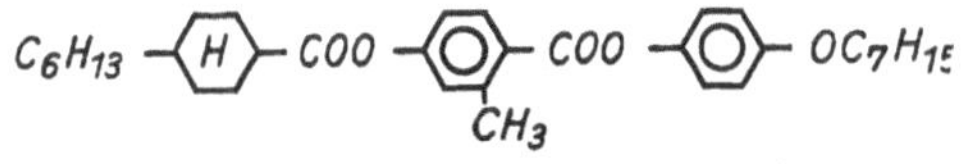

K 56 N 133 Is (DEUTSCHER et al.)

Smectogenic components:

C:

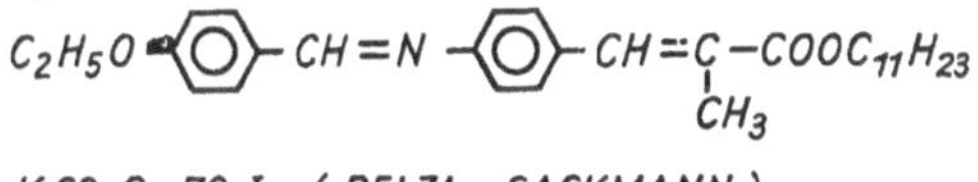

K 62 S_A 78 Is (PELZL, SACKMANN)

D:

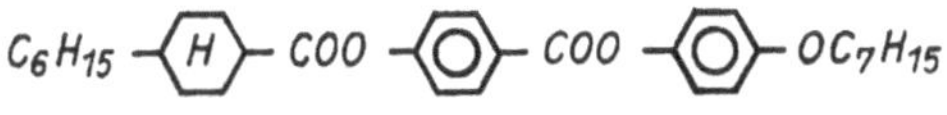

K 64 S_B 94 S_A 168 N 192 Is (DEUTSCHER et al.)

[1]) Contributions I, II, III are given by PELZL, LATIF et al., PELZL, SCHOLZ et al., PELZL, SCHOLZ, DEMUS, SACKMANN).

K: solid crystal
N: nematic
S_A, S_B: smectic A resp. B
Is: isotropic liquid

The numbers between the phase symbols designate the transition temperatures (°C)

The phase diagrams were studied by microscopical observation of the transition temperatures of samples made up at specific concentrations.

3. Experimental results

The phase diagram of the binary system A/C is shown in Figure 1. In this diagram a wide region of an S_A mixed phase is observed which shows a pronounced maximum in the transition curve $S_A \rightarrow N$. The N_{re} phase appears in a relatively wide concen-

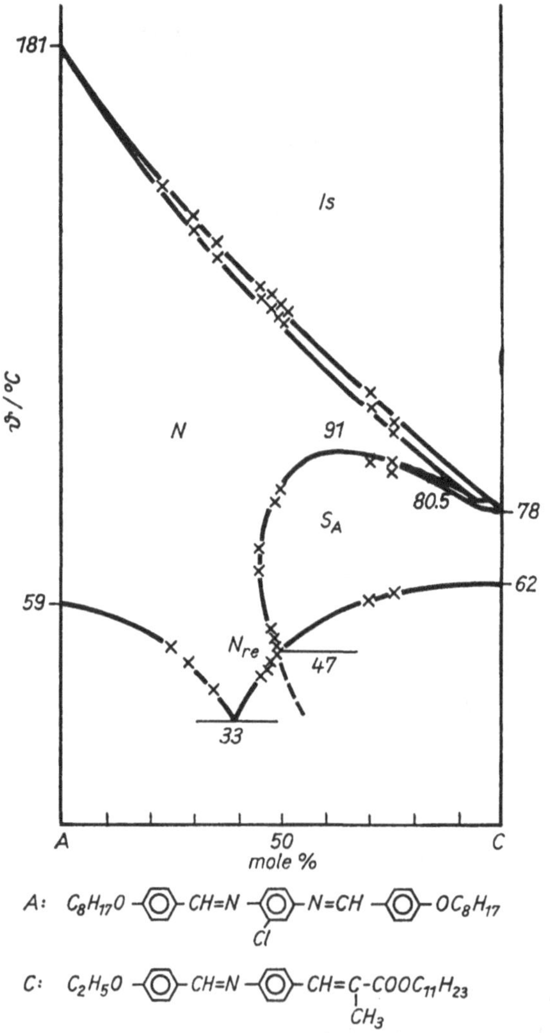

Fig. 1. Phase diagram of the binary system with the components A and C

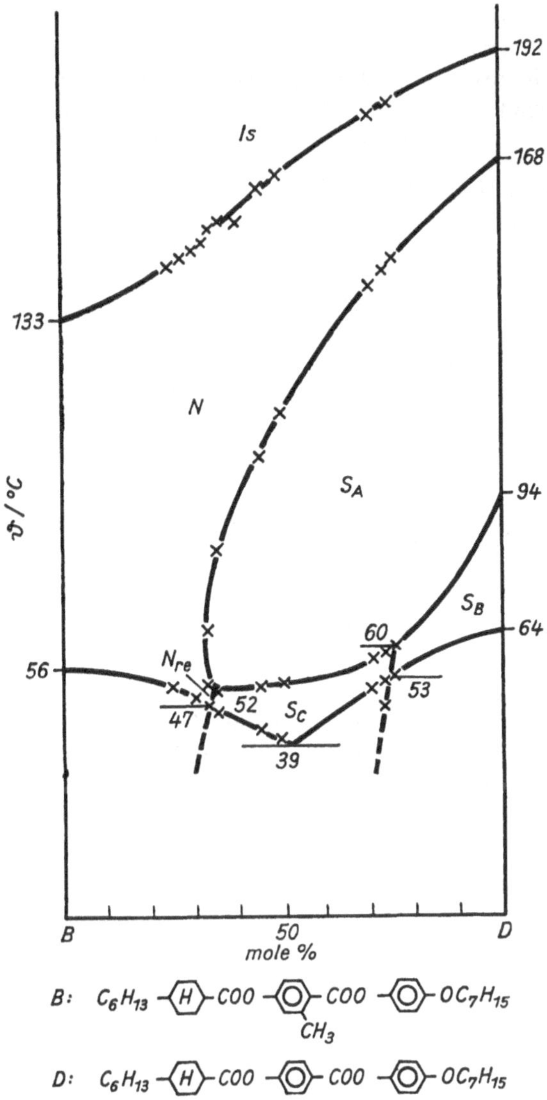

Fig. 2. Phase diagram of the binary system with the components B and D

tration range between 45 and 55 mole% A; between 45 and 50 mole% A this N_{re} phase is a metastable phase[2]).

As seen in Figure 2, the phase diagram of the binary system B/D is more complicated. This is due to the fact that the smectogenic component exhibit two smectic phases (S_A, S_B) and that for middle concentrations an intermediate S_C phase occurs. The N_{re} phase exists only in a very small concentration range (33 mole% B).

4. Discussion

As mentioned before a minimum length of the alkyl chains of the nematogenic as well as the smectogenic components is an important condition for the formation of

[2]) Phase transitions on the supercooled state were drawn by dotted lines.

N_{re} phases in binary systems of terminal-nonpolar compounds (Pelzl, Scholz, Demus, Sackmann). Up to now only asymmetrical nematogenic compounds are used. This seems to be not accidental because for symmetrical compounds with sufficient long alkyl chains the tendency to form smectic phases dominates. On the other hand, in compounds with two long terminal alkyl chains the smectic properties can be suppressed by lateral substituents (see compounds A and B). It follows from our studies that such terminal-nonpolar compounds are quite suited as nematogenic components in binary systems with N_{re} phases.

References

Deutscher, H.-J., Seidel, C., Körber, M., Schubert, H.: J. prakt. Chem. **321** (1979) 47

Neubert, M. E., Maurer, L. J.: Mol. Cryst. Liq. Cryst. **43** (1977) 313

Pelzl, G., Diele, S., Latif, I., Weissflog, W., Demus, D.: Cryst. Res. Technol. **17** (1982) K78

Pelzl, G., Latif, I., Diele, S., Novak, M., Demus, D., Sachmann, H.: Mol. Cryst. Liq-Cryst. **139** (1986) 333

Pelzl, G., Scholz, C., Demus, D., Sackmann, H.: (in preparation)

Pelzl, G., Scholz, C., Diele, S., Deutscher, H.-J., Demus, D., Sackmann, H.: (in preparation)

Pelzl, G., Sackmann, H.: Symposium Farad. Soc. No. 5 (1971) 68

(Received October 12, 1987)

Authors' address:

Dr. sc. G. Pelzl, Dipl.-Chem. C. Scholz, Prof. Dr. habil. H. Sackmann
Sektion Chemie der Martin-Luther-Universität Halle-Wittenberg
Mühlpforte 1
DDR-4020 Halle/Saale

Crystal Research and Technology

Zeitschrift für experimentelle und technische Kristallographie

Edited by H. Neels, Leipzig

Papers in Number 3, 1988

Original Papers

AKADEMIE-VERLAG · BERLIN

Continued from cover page 4

Short Notes

Book Reviews

Crystal Research and Technology
Volume 23 1988 Number 2

Contents

Original Papers

Continued from cover page 3

Crystal Research and Technology is indexed in Current Contents/PC & ES.

FSC
www.fsc.org
MIX
Papier | Fördert
gute Waldnutzung
FSC® C083411

Kolibri 360 GmbH
Ferdinand-Jühlke-Straße 7
99095 Erfurt, Deutschland
produktsicherheit@kolibri360.de